Weed Control Methods for Rights-of-Way Management

Editor

Edward O. Gangstad
Aquatic Plant Control Specialist
U.S. Army Corps of Engineers
Washington, D.C.

CRC Press, Inc.
Boca Raton, Florida

Library of Congress Cataloging in Publication Data

Main entry under title:

Weed control methods for rights-of-way
management.

Includes bibliographies and index.
1. Weed control. 2. Right of way.
3. Herbicides. 4. Herbicides — Environmental
aspects. 5. Weed control — Law and legis-
lation — United States. I. Gangstad, Edward O.
SB611.W36 628.9'7 81-7743
ISBN 0-8493-5329-7 AACR2

Direct all inquiries to CRC Press, Inc., 2000 N.W. 24th Street, Boca Raton, Florida, 33431.

International Standard Book Number 0-8493-5329-7

Library of Congress Card Number 81-7743
Printed in the United States

PREFACE

Suppression and/or eradication of unwanted vegetation on or along rights-of-way has been a major phase of chemical weed control. Until recent times, mechanical methods were the only methods available for controlling this vegetation. The use of herbicides in combination with mechanical, cultural, and biological methods of control has greatly expanded the capability to effectively and economically control vegetation in these areas. This volume describes in detail methods of control and related data including, (1) all vegetation on industrial sites, paved highways, and railroad ballast, (2) woody vegetation along roadsides, utility lines, and fire breaks, and (3) aquatic vegetation on rivers and stream banks, waterways, ponds, reservoirs, irrigation, and drainage channels.

Edward O. Gangstad

THE EDITOR

Dr. E. O. Gangstad, Aquatic Plant Control Specialist, transferred to the Office of the Chief of Engineers, Washington, D.C. in October of 1966 from the Texas Research Foundation, Dallas, where he was employed as Principal Agronomist of the Hoblitzelle Agricultural Laboratory. Dr. Gangstad was graduated from the University of Wisconsin, Madison, with an M.A. in Biochemistry (1947) and from Rutgers University, New Brunswick, N.J., with a Ph.D. in Agronomy (1950) and minors in Plant Pathology and Plant Physiology.

During the past 25 years he has established an outstanding career in the environmental and biological fields and has over 80 scientific publications. He is a contributor to the *Journal of Biological Chemistry, Agronomy Journal, Crop Science, Journal of Heredity, Weed Science, Crops and Soils, Tropical Agriculture, Turrialba, Soil Science Society of Florida, Hoblitzelle Agricultural Bulletin, Wrightia, Hyacinth Control Journal, Environmental Quality Journal,* and the *Journals of the American Society of Civil Engineers.*

Dr. Gangstad's experience includes research and field studies on the physiology and pathology of plants at the Universities of Wisconsin and New Jersey; research and development studies on strategic long-vegetable fibers and related programs in Florida with the U.S. Department of Agriculture, Office of Naval Research, and Department of Defense; and agrobusiness studies to develop renewable natural resources of the State of Texas with the Texas Research Foundation and management of natural resources for the Department of the Army.

Dr. Gangstad is listed in *American Men of Science, Leaders in Science, American Men of Achievement,* and the *International Who's Who of Intellectuals.*

IN MEMORIUM

Thomas Traquair McClure

The death of Thomas Traquair McClure on July 21, 1973 ended the career of a distinguished plant pathologist. Born in Boston on August 30, 1914, he began his studies in plant pathology in 1939 at Iowa State after finishing a bachelors of science in biology at Harvard in 1936 and a masters in plant physiology at Rutgers in 1938. His studies at Iowa State were disrupted by military service in the Army from 1942 to 1946. He returned to his studies at the University of California at Berkeley and was granted a doctorate in Plant Pathology in 1949 by the University of California.

He had been an active member in the Society of Sigma Xi, The American Phytopathological Society, the American Society of Plant Physiologists, the American Association for the Advancement of Science, the Botanical Society of America, the Weed Science Society of America, and was chairman of the committee on Aquatic Plant Control, (including Herbicides and Algacides) of the American Society of Testing Methods and Materials.

Those who had the good fortune of knowing Tom will remember him for his quick wit and humor. His philosophy of being a scientist was inspiring to many.

CONTRIBUTORS

C. Grant Ash (Retired)
Environmental Policy
Civil Works
U.S. Army Corps of Engineers
Washington, D.C.

L. L. Danielson (Retired)
Agricultural Research Service
U.S. Department of Agriculture
Beltsville, Maryland

Wendell R. Mullison (Retired)
Registration Specialist
Health and Environmental Research Department
The Dow Chemical Company
Midland, Michigan

Philip J. Phillips (Retired)
North Atlantic Division
U.S. Army Corps of Engineers
New York, New York

C. M. Schwartz (Deceased)
Natural Resource Specialist
Office of the Chief of Engineers
Washington, D.C.

TABLE OF CONTENTS

Part I: Basic Concerns Of Weed Control Management

Chapter 1

THE SCIENCE OF WEED CONTROL*

INTRODUCTION

Weed science includes the study of weeds and their control by cultural, mechanical, biological, and chemical means. Weed scientists are engaged in varied activities, including research, teaching, extension, regulation, product development, and marketing. They are involved in controlling weeds in field crops, horticultural crops, pastures and rangeland, forests, aquatic areas, and noncropland, such as lawns and recreation areas, roadways, railroad, and utility rights-of-way. These scientists in industry, universities, and government are continually developing new chemical compounds and new methods of control. We need to intensify our national effort to understand and weigh the immediate and long-term effects of current and new methods of weed control.

Prior to the advent of the phenoxy herbicides in 1944, the control of weeds involved costly and time-consuming mechanical methods supplemented by hand labor. The discovery of 2,4-dichlorophenoxy acetic acid (2,4-D) and related compounds initiated an intensive research effort over the years, resulting in the development of numerous additional herbicides of value in right-of-way weed-control programs.

Herbicides have proved to be efficient tools to use independently or in combination with mechanical methods for suppressing or eradicating unwanted vegetation on noncrop land, including areas along paved highways and railroads; woody plants in grassland on utility rights-of-way; and marginal and aquatic weeds in drainage ditches and watersheds, waterways, ponds, lakes, and reservoirs. Special attention is drawn to the term "unwanted vegetation" to emphasize the point that the undesirable species are selectively removed by careful planning in the choice of herbicides and times of application in control programs. As a result, desirable wild plant species remain to thrive and minimize erosion, preserve plant nutrients, and provide food and cover for wildlife.

HERBICIDE HAZARDS

The hazards of herbicides to the applicator and the handler and the dangers of these materials to livestock, fish, and wildlife are crucial concerns in developing acceptable and successful weed-control programs. Potential injury to desirable plants and the requirements for effective performance are additional considerations of great importance in selecting herbicides, dosages, and times of application for specific problems.

An understanding of the biology of weed species is necessary for the accurate selection of appropriate herbicides, dosage levels, and times and frequency of application in planning combination methods of control. A knowledge of the physical, chemical, and biological properties of herbicides is essential in using them safely and efficiently.

Any hazards involved in the use of herbicides are avoided or minimized by the careful observance of all precautions and directions for safe use and storage given on the manufacturer's label. Before a herbicide can be sold on the open market, the label must be approved by the U.S. Environmental Protection Agency (EPA). Label approval is based on information provided by research, development, and regulatory personnel of state, federal, and industry organizations.[1-3,5,7-10]

* By L. L. Danielson (retired), formerly of the U.S. Department of Agriculture, Agricultural Research Service, Beltsville, Maryland.

CLASSIFICATION OF WEEDS

Weeds are grouped as annuals, biennials, or perennials according to their longevity and the time required to complete their life cycles. The many species of plants represented in these groups is an indication of the need for carefully planned programs of control as opposed to repeated individual emergency measures.

Annual weeds mature in one season and are propagated by seeds. Some examples of representative annual weeds are crabgrass (*Digitaria* spp.), foxtail (*Setaria* spp.), pigweed (*Amaranthus* spp.), common ragweed (*Ambrosia artemisifolis* L.), wild buckwheat (*Polygonum convolvulus*), and cocklebur (*Xanthium* spp.). These and similar weeds that emerge and mature during warm weather are commonly called summer annuals. Winter annuals, such as pennycress (*Thlaspi arvense* L.), common chickweed (*Stellaria media* (L.) Cyrillo), shepherd's purse (*Capsella bursa-pastoris* (L.) Medic.), and annual bluegrass (*Poa annua* L.), germinate in the fall, live in the rosette stage through the winter, and mature in the spring or early summer. The capacity of individual annual weed plants to produce thousands or even hundreds of thousands of seeds is a means of ensuring the survival of each species. For example, a field dodder plant (*Cuscuta* spp.) may produce more than 16,000 seeds; a pigweed plant may produce more than 100,000; and a lamb's-quarters (*Chenopodium album* L.) plant may produce more than 70,000 seeds.[12] The seeds of many species remain viable for many years when buried in the soil.[4,11] Seeds of some common weeds have survived for more than 80 years.[(6)] This is additional insurance for the survival of each species. These properties of annual weeds make eradication of them almost impossible. Eventual reduction of populations can be achieved by killing annual weeds before maturation to prevent seed formation. Destruction of the tops and all axillary and terminal buds prevents regrowth and seed production of annuals and breaks the life cycle of the current population. New plants will emerge from the seed reservoir in the soil in the following and successive years. A continuing program is required to control annual weeds.

Biennials require two seasons to complete their life cycles. There are very few biennials in the temperate regions. These plants live for more than 1 year, but not more than 2 years. Because they reproduce from seeds only, the control measures are in general the same as for annuals. Examples of biennials are wild carrot (*Daucus carota* L.), bull thistle (*Cirsium vulgare* (Savi) Tenore), burdock (*Arctium* spp.), and common mullein (*Verbascum* spp.).

Perennial weeds live more than 2 years. They reproduce from seeds and have the additional capability of reproducing vegetatively. They have storage organs, such as stolons (prostrate stems), rhizomes (underground stems), bulbs, crowns, and roots. Carbohydrates and other nutrients are stored in these organs to nourish the new growth in the beginning of each growing season. The new growth comes from a bud and is sustained on the stored food until it becomes established. Unlike annual plants, the top growth of perennials can be destroyed without killing them due to their ability to regenerate new growth from storage organs. To kill perennials, it is necessary to restrict accumulation of food reserves or destroy the storage organs. Photosynthesis is the basic source of food reserves. Reserve accumulation can be reduced or prevented by restricting photosynthesis by chemical means or by repeated removal of the tops of perennial plants. Storage reserves can be depleted by allowing new growth to become established repeatedly and destroying it. Example of perennial weeds are quackgrass (*Agropyron repens* (L.) Beauv.), Canada thistle (*Cirsium arvense* (L.) Scop.), and nutsedge (*Cyperus* spp.).

CLASSIFICATION OF HERBICIDES

Weed species respond differentially to herbicides. Crops and other valuable plants show similar responses. Based on these effects, herbicides are termed selective or nonselective forms. For convenience, they may be categorized on the basis of mode of action as contact, translocated, sterilant, and preemergence chemicals.

Selective herbicides can be used to kill certain weed species without seriously injuring plantings of desirable species infested by them. Those that kill crabgrass or dandelions (*Taraxicum* spp.) in grass sod are good examples. The selectivity of herbicides in some mixtures of weeds and desirable plants is known; in other combinations of plants, they are unknown. For example, crabgrass can be killed in bluegrass sod because it is an annual growing among perennial plants that can recover from spray injury.

Nonselective herbicides kill vegetation with little discrimination. Certain weed species, however, are resistant, and some escape. Resistant species are physiologically resistant to the herbicide; some escapees are perennials that have part of their root systems below treated layers of soil; others are annuals and shallow-rooted perennials that reinfest an area after the chemical has leached below the surface layer.

Contact herbicides, such as petroleum oils, sulfuric acid, and the dinitro compounds, kill tissues that are wetted with the spray. Whether the plant dies or recovers depends on whether it has a protected growing point. Perennials usually have underground buds that will regrow.

Translocated chemicals are absorbed by the leaves and stems or by the roots and move through the vascular system to leaves, buds, and root tips. When absorbed by the leaves and stems, the chemical is commonly moved with the food materials that are manufactured in the leaves and stems. When absorbed by the roots, it moves in the water-conducting tissue. The growth-regulator type of translocated herbicide is a synthetic compound that functions like a plant hormone. It accumulates mostly in areas of rapidly dividing cells, upsetting the normal metabolism of the plant and causing death of the cells. Foliar applications of translocated herbicides are of great practical use, as small amounts are effectvie and they can be applied in small volumes of water or oil.

Soil sterilant herbicides prevent the growth of all higher plant life. These include sodium arsenite, sodium chlorate, sodium metaborate tetrahydrate, and others. These chemicals do not necessarily kill all life in the soil, such as fungi, bacteria, and other microorganisms. Their toxic effects may remain for only a short time or for years. Residual toxicity depends on the chemical and its rate of decomposition or leaching, the colloidal and chemical content of the soil, species tolerance, and rate of application.

1. Herbicides vary in their rate of disappearance from the soil because of volatility, susceptibility to decomposition by soil microorganisms, and solubility. For example, some of the carbamates are volatile at high temperatures and rapidly lose their toxic effect during the summer months. Certain soil microorganisms effectivly decompose 2,4-D.
2. Some herbicides are readily adsorbed by mineral and organic colloids and are rendered unavailable or made slowly available to plants. The fertility and pH of soils are also influencing factors in the persistence or availability of toxic amounts of certain chemicals. For example, monuron [3-(*p*-chlorophenyl)-1,1-dimethylurea] and diuron [3-(3,4-dichlorophenyl)-1,1-dimethylurea] are adsorbed on clay colloid particles so that leaching is difficult.

3. Plants species vary widely in tolerance to soil sterilants. Deep-rooted woody plants are generally more tolerant to soil sterilants than succulent, shallow-rooted plants.
4. High rates of application generally last longer than low rates.

Preemergence herbicides, such as certain carbamates, nitroanilines, triazines, and others, are those applied to the soil before the foliage of the weeds emerge from the soil. With reference to crops, the treatment may be made before planting, at planting, or after planting, but before emergence. Some may also be used following clean cultivation in established crops. These herbicides kill by contact or may be translocated from the point of entry into roots, stems, or leaves. Some inhibit photosynthesis, while others affect growth processes such as cell division and elongation. The amount of residual action in the soil is directly related to dosage level and the biological and chemical properties of the herbicides.

HAZARDS OF HANDLING AND APPLICATION

Nearly all herbicides are potentially dangerous in one way or another, but they are not likely to cause injury if used properly and if recommended precautions are observed.[1,9,14] Several kinds of dangers are associated with the handling and application of these chemicals, and possible injury is not limited to the operator. Any or all of the following may be affected: operator and handler, livestock, desirable plants, equipment, and fish and wildlife.

The following general discussion of the possible hazards involved in the handling and application of herbicides applies to classes of herbicides rather than individual herbicides. The operator and handler must carefully read the label on each container before using or storing the product. Follow all instructions and heed all precautions and warnings, giving special attention to the antidote statement for each chemical. Store in original labeled containers. Dispose of empty containers by following the most recent directions for disposal supplied by the EPA. These are available from the state pesticide regulatory agencies.

The operator or handler who mixes and applies the spray or spreads the dry product could be poisoned from swallowing the herbicide, from skin absorption, or from inhalation. In each instance, there is greater danger from the concentrated material than from the diluted spray solution or suspension. Some general principles are helpful in handling emergency situations.

1. If concentrated spray is ingested, immediate and appropriate action should be taken, depending on the chemical. In most instances, vomiting should be induced by giving the victim a tablespoon of salt in a glass of warm water. Repeat the treatment until the fluid is clear. Keep the patient quiet and warm until a physician arrives. Do not attempt to give fluids by mouth or to induce vomiting in an unconscious person. With some herbicides, such as endothal [7-oxabicyclo(2,2,1) heptane-2,3-dicarboxylic acid], do not induce vomiting, but give the patient large quantities of milk or whites of eggs beaten in water. Be sure to read the label on the herbicide container to be certain that the treatment given is correct. Again, reading the label carefully before use of a herbicide should be a regular procedure. The acute toxicity of herbicides varies widely as taken directly. Toxicity in the field is usually less because seldom is such a dose swallowed or absorbed, and diluted sprays are handled more often than concentrates. Nevertheless, neither chemical concentrates nor diluted sprays should be kept in unlabeled containers, especially not in containers commonly used for potable liquids. Unfor-

tunately humans do not always react to chemicals in the same way as small animals do. Experience has shown that it is always possible that a human being will tolerate a correspondingly larger dose of a given chemical than results with animals indicate or vice versa. There is no way to determine accurately whether a lethal dose for man is larger or smaller than the LD_{50} (lethal dose for 50% of animals tested) for a laboratory animal. However, animal studies are the best guide available.

2. Absorption by the skin and irritation of skin and eyes can usually be prevented. Keep exposure to a minimum. Some individuals are hypersensitive to certain chemicals and have allergic reactions that are impossible to predict without skin tests. For most herbicides, washing hands and face with soap and water after handling is sufficient protection. Prolonged contact is more dangerous than short exposures. For the more readily absorbed chemicals and those that are irritating, wear clean clothing that covers the body. Remove clothing after it has become contaminated with the chemical. Use synthetic rubber gloves. Where splashing may occur, wear goggles. If spray or dust is spilled on the skin, wash thoroughly with soap and water; if spray or dust is spilled into the eyes, flush with plain water and see a doctor. Liquid concentrates and powders should be put into containers that can be readily lifted by the operator in the field. Packages of powders should be small enough so that it is unnecessary to remove the contents with a scoop.
3. Inhaling vapors, dusts, and spray mists can also be avoided. Use a mask approved for the particular type of exposure, when label directions indicate the need. In the situation where a severe exposure has occurred, move the patient into fresh air, administer artificial respiration if needed, and call a physician.
4. Some chemicals are flammable or support fire. Avoid ignition from friction, sparks, and contact with combustible materials.
5. Some are dyes that color skin and hair if appropriate protection is not provided.

The chief dangers of poisoning livestock when using most herbicides result from animal comsumption of herbicide remnants in open containers or spilled on the ground or floor. Very toxic herbicides, such as sodium arsenite, kill livestock that eat the treated forage.

Herbicides are often used to eradicate or control undesirable plants that are poor animal forage. Therefore, the livestock environment is usually improved through chemical control by the replacement of weeds with desirable forage plants. Some general precautions must be observed because of the specific effects of herbicides on plants. For example, 2,4-D and similar compounds may increase the palatibility of plants not ordinarily eaten. If these are poisonous species, such as jimson weed (*Datura stramonium* L.), larkspur (*Delphinium* spp.), and hemlock (*Conium maculatum* L.), sickness and death may result. The nitrate content of some plants may be increased when sprayed with 2,4-D. This nitrate is reduced to nitrite by microorganisms in the intestinal tract of herbivorous animals. Because nitrite in the bloodstream interferes with the effective transport and use of oxygen, the animal suffocates. However, increased nitrate content of plants does not always follow spraying. Furthermore, some natural conditions also produce high nitrates, so that many weeds contain enough to cause poisoning whether sprayed or not. More commonly, nonfertilized soils are so low in nitrogen content that nitrite poisoning is no hazard.

The prevention of damage to valuable plants adjacent to treated areas requires special precautions. This damage may be the result of drifting, washing, or leaching of the herbicide.

1. Drift hazards are greatest when herbicides affecting the leaves of plants are used. These may be of the growth-regulating type, such as 2,4-D, 2,4,5-trichlorophenoxy (2,4,5-T) acetic acid, and silvex [2-(2,4,5-trichlorophenoxy)propionic acid], or of the contact type, such as pentachlorophenol (PCP), the petroleum oils, and dinitros. Danger is minimal when granular formulations of nonvolatile herbicides are used. Drift occurs not only with volatile herbicides, such as the high-volatile esters of 2,4-D and 2,4,5-T, but also from sprays that have been atomized into mists by high pressure and small nozzle opening whether the formulations are volatile or not.
2. Washing of treated soils is a hazard on slopes, bare soil surfaces, and pavements. Herbicides are often carried by surface runoff to valuable plants downslope. Accordingly, it is most important that equipment is not drained or flushed where runoff to desirable plants may occur.
3. Leaching moves chemicals downward through the soil. If they are readily absorbed by the roots, the roots of plants that extend under the treated area are likely to be injured. Treatment of such areas with soil sterilants should be avoided. Application equipment should not be drained or flushed where leaching to the roots of desirable plants may occur.

Spraying equipment can be severely damaged by the corrosion of metal parts due to exposure to chemicals. Petroleum oils and other solvents destroy rubber gaskets, liners, and hoses. Thorough draining and cleaning of equipment with water and a detergent is sufficient protection against most chemicals. When corrosive chemicals are used, coat the metal parts of the equipment with protective paints, oil, or undercoating after use or purchase equipment with noncorrosive metals. Teflon® and neoprene rubber are resistant to oils and solvents.

Wildlife may suffer primary or secondary effects from herbicide usage. The primary effects of herbicides on wildlife have not been serious.

Secondary effects of herbicides on wildlife include toxicological effect due to sublethal doses, changes in palatability of poisonous plants, changes in the chemical composition of plants, and ecological effects in the habitat by injury, contamination, or death of organisms in the food chain. All herbicides have the potential of exerting major secondary effects on wildlife through ecological change because they are used to eliminate or reduce the numbers of certain plants to provide for the increase of other plants. These ecological changes in the habitat may well lead to increases or decreases in wildlife numbers.

There are a few herbicides, such as the dinitros, that can poison animals. Most injury results from overdoses and spillage. Promiscuous spraying can destroy cover, but herbicides can also be useful in management. Openings in wooded areas, such as the clearing for utility company rights-of-way and spraying of hardwoods in stands of pines, can be beneficial to wildlife.

A few herbicides are very toxic to fish, but many can be used for the control of aquatic weeds. The control of submersed weeds in ponds or streams can be beneficial to fish populations. Safe amounts of herbicides, expressed in parts of the chemical per million parts of water, vary widely with age, size, and species of fish involved.

Labels for commercial herbicide containers are written with great care to insure that only facts are stated. Labels for all pesticides must be approved by the EPA before they are used. The applicator should read the entire label before using any herbicide. If the directions and precautions for use and storage are not clear, the dealer and/or a representative of the state pesticide regulatory agency should be contacted for clarification.

Labels give the exact name of the active ingredient. For example, 2,4-D is sold in several different forms, such as the sodium or amine salt or a volatile or low-volatile ester. Specific recommendations are given for the use of each form or formulation of the same basic chemical because of their differing effects.

The labels state the amount of acid equivalent, phenol equivalent, or active ingredient in the product. This information facilitates a comparison of concentrations in various formulations, a useful procedure for determining the amounts to apply and for contrasting prices. The label also give recommendations for use and rates and times of application. Certain warnings are stated, when necessary, to protect the operator and/or consumer from any possible hazardous effects or irritation by the chemical and to protect desirable plants from injury.

WARNINGS AND CAUTION STATEMENTS ON THE LABEL

Required warnings and caution statements concerning the general areas of toxicological hazard, including hazard to children, environmental hazard, and physical or chemical hazard, fall into two groups: those required on the front panel of the labeling and those which may appear elsewhere.[13,14]

Specific requirements concerning content, placement, type size, and prominence are outlined below.

Required Front Panel Statements

With the exception of the child hazard warning statement, the text required on the front panel of the label is determined by the toxicity category of the pesticide assigned according to the table below:

	Toxicity categories			
Effects	I	II	III	IV
Oral LD_{50}	50 mg/kg	500 mg/kg	5,000 mg/kg	5,000 mg/kg
Inhalation LC_{50}				
Dust or mist	2.0 mg/ℓ	20	200	200
Gas or vapor	200 ppm	2,000	20,000	20,000
Dermal LC_{50}	200 mg/kg	2,000	20,000	20,000
Eye effects	Irreversible corneal opacity at 7 days	Corneal opacity reversible within 7 days or irritation persisting for 7 days	No corneal opacity irritation reversible within 7 days	No irritation
Skin irritation	Severe irritation or damage at 72 hr	Moderate irritation at 72 hr	Mild or slight irritation at 72 hr	No irritation at 72 hr

Human Hazard Signal Word

Toxicity Category I — All products falling into Toxicity Category I bear the signal word "Danger". In addition, if the pesticide was assigned to Toxicity Category I on the basis of its oral, inhalation, or dermal toxicity (as distinct from skin and eye effects), the word "Poison" appears in red on a background of distinctly contrasting color, and the skull and crossbones shall appear in immediate proximity to the word "poison".

Toxicity Category II — All products falling into Toxicity Category II bear the signal word "Warning".

Toxicity Category III — All products falling into Toxicity Category III bear the signal word "Caution".

Toxicity Category IV — No signal word is required for products falling in Toxicity Category IV.

Use of Signal Words — The use of the signal word(s) associated with the toxicity category into which the product falls is mandatory. Use of the signal word(s) associated with a higher toxicity category is not permitted except when in the opinion of EPA such overlabeling is necessary to prevent unreasonable adverse effects on man or the environment. No more than one human hazard signal word is allowed to appear on the front panel of a label.

Child Hazard Warning

Every pesticide label bears on the front panel the statement "Keep Out of Reach of Children". Only in cases where the likelihood of contact with children during distribution, marketing, storage or use is demonstrated by the applicant to be extremely remote may EPA waive this requirement.

Statement of Practical Treatment

Toxicity Category I — A statement of practical treatment (first aid or other) appears on the front panel of the label of all pesticides falling into Toxicity Category I. EPA may, however, permit reasonable variations in the placement of the statement of practical treatment if some reference such as "See statement of practical treatment on back panel" appears on the front panel near the word "Poison" and the skull and crossbones.

Lower Toxicity Categories — The statement of practical treatment is not required for products falling into Toxicity Categories II, III, and IV.

Placement and Prominence

All the required front panel warning statements are grouped together on the label and appear with sufficient prominence relative to other front panel text and graphic material to make them unlikely to be overlooked under customary conditions of purchase and use. The following table shows the minimum type size requirements for the front panel warning statements on various-sized labels.

Size of label front panel ($in.^2$)	Required signal word, all capitals	"Keep Out of Reach of Children"
5 and under	6 point	6 point
Above 5 to 10	10 point	8 point
Above 10 to 15	12 point	8 point
Above 15 to 30	14 point	10 point
Over 30	18 point	12 point

Other Required Warnings and Caution Statements

The warnings and caution statements shall appear together on the label under the general heading "Caution Statements" and under appropriate subheadings of "Hazard to Humans and Domestic Animals", "Environmental Hazard", and "Physical or Chemical Hazard".

Hazard to Humans and Domestic Animals

For each indicator of hazard included in the table of toxicity categories above, this section shall include a clear statement of the particular hazard and the route(s) of exposure and a statement of the precautions to be taken to avoid accident, injury, or damage.

The following table describes required statements. These statements may be modified to reflect specific hazards and should be considered exemplary rather than definitive.

Required Caution Statements by Toxicity Category

Toxicity category	Oral, inhalation, or dermal toxicity	Skin and eye irritation
I	Poisonous if swallowed (inhaled or absorbed through skin); do not breathe vapor (dust or spray mist); do not get in eyes, on skin, or on clothing	Corrosive, causes eye and skin damage and skin damage and skin irritation; do not get in eyes, on skin, or on clothing; wear goggles or face shield and rubber gloves when handling; harmful or fatal if swallowed
II	May be fatal if swallowed (inhaled or absorbed through the skin); do not breathe vapors (dust or spray mist); do not get in eyes, on skin, or on clothing	Causes eye (and skin) irritation; do not get in eyes, on skin, or on clothing; harmful if swallowed
III	Harmful if swallowed (inhaled or absorbed through the skin); avoid breathing vapors (dust or spray mist); avoid contact with skin (eyes or clothing).	Avoid contact with skin, eyes or clothing; first aid: in case of contact immediately flush eyes with plenty of water; get medical attention if irritation persists
IV	No caution statement required	No caution statement required

Environmental Hazards

Where a hazard exists to birds, fish, wildlife, or pollinating insects, caution statements are required stating the nature of the hazard and the necessary precautions to avoid accident, injury, or damage to nontarget species or to the environment. Examples of such caution statements and the circumstances under which they are required follow.

1. If a pesticide for outdoor use has a mammalian acute oral LD_{50} of 100 mg/kg or less, the statement "Toxic to Wildlife" is required.
2. If a pesticide for outdoor use has a fish acute LC_{50} of 1 ppm or less, the statement "Toxic to Fish" is required.
3. If a pesticide for outdoor use has an avian acute oral LD_{50} of 100 mg/kg or less, or a subacute dietary LC_{50} of 500 ppm or less, the statement "Toxic to Birds" is required.
4. If either accident history or field studies indicate any kill of birds or mammals resulting from use of the pesticide, the statement "Birds and Other Wildlife in Treated Areas May Be Killed" is required.
5. For uses involving foliar application to agricultural crops, forests, or shade trees or for mosquito abatement treatments, pesticides toxic to bees must bear appropriate label cautions.
6. For all outdoor uses other than aquatic applications, the label must bear the caution "Keep out of Lakes, Ponds, or Streams; Do Not Contaminate Water by Cleaning of Equipment or Disposal of Wastes".
7. Appropriate instructions and cautions concerning safe disposal of used pesticide containers may be required.

Physical or Chemical Hazards

Warnings concerning flammability, explosive, or other physical hazards are required in accord with the physical properties of the pesticide. Flammability cautions and the circumstances under which they are required are as follows:

Pressurized Containers

Characteristics	Required text
Flash point at or below 20°F; or if there is a flashback at any valve opening	Danger: extremely flammable. Contents under pressure. Keep away from fire, sparks and heated surfaces. Do not puncture or incinerate container. Exposure to temperature above 130°F may cause bursting.
Flash point above 20°F and not over 80°F; or if the flame extension is more than 18 in. long at a distance of 6 in. from the flame	Warning: flammable. Contents under pressure. Keep away from heat, sparks, and open flame. Do not puncture or incinerate container. Exposure to temperatures above 130°F may cause bursting.
All other pressurized containers	Warning: Contents under pressure. Do not use or store near heat or open flame. Do not puncture or incinerate container. Exposure to temperatures above 130°F may cause bursting.

Nonpressurized Containers

Flash point	Required text
At or below 20°F	Danger: extremely flammable. Keep away from fire, sparks and heated surfaces.
Above 20 and not over 80°F	Warning: flammable. Keep away from heat and open flame.
Above 80 and not over 150°F	Caution: do not use or store near heat or open flame.

Directions for Use — General Requirement

Adequacy and Clarity of Directions

Directions for use must be stated in terms which can be easily read and understood by the average person likely to use or to supervise the use of the pesticide. When followed, directions must be adequate to protect the public from economic cheat and from personal injury and to prevent unreasonable adverse effects on the environment.

Placement of Directions for Use

Directions may appear on any portion of the label provided that they are conspicuous enough to be easily read by the user of the pesticide. Directions for use may appear on printed or graphic matter which accompanies the pesticide, provided that:

1. If required by the EPA, such printed or graphic matter is securely attached to each package of the pesticide or placed within the outside wrapper or bag.
2. The label bears reference to the directions for use on accompanying leaflets or circulars such as "See directions on the enclosed circular."
3. In the opinion of EPA, it is not necessary for such directions to appear on the label.

Exceptions to Requirement for Directions for Use

Detailed directions for use may be omitted from labeling of pesticides which are intended for use only by manufacturers of products other than pesticides in their regular manufacturing processes, provided that:

1. The label clearly shows that the product is intended for use only in manufacturing processes and specifies the type(s) of products involved.

2. Adequate information, such as technical data sheets or bulletins, is available to the trade specifying the type of pesticide involved and its proper use in manufacturing processes.
3. The pesticide will not come into the hands of the general public except after incorporation into such finished products.
4. In the opinion of EPA, such directions are not necessary for the protection of health and the environment.

Detailed directions for use may be omitted from the labeling of pesticides for which sale is limited to physicians, veterinarians, or druggists, provided that:

1. The label clearly states that the product is for use only by physicians or veterinarians.
2. In the opinion of EPA, such directions are not necessary for the protection of health and the environment.
3. The product is also a drug.

Detailed directions for use may be omitted from the labeling of pesticides which are intended for use only by formulators in preparing pesticides for sale to the public, provided that:

1. There is information readily available to the formulator on the composition, toxicity, methods of use, applicable restrictions or limitations, and effectiveness of the product for pesticide purposes.
2. The label clearly states that the product is intended for use only in manufacturing, formulating, mixing, or repacking for use as pesticides and specifies the type(s) of pesticide involved.
3. In the opinion of EPA, such directions are not necessary for the protection of health and the environment.

Contents of the Label

The directions for use must include the following under the heading "Directions for Use":

1. The statement of use classification immediately under the heading "Directions for Use"
2. Immediately below the statement of use classification, the statement: "It is a violation of Federal law to use this product in a manner inconsistent with its labeling."
3. The site of application, as for example, the crops, animals, areas, or objects to be treated
4. The target pest(s) associated with each site
5. The dosage rate associated with each site
6. The method of application, including instructions for dilution, if required, and type(s) of application apparatus or equipment required
7. The frequency and timing of applications necessary to obtain effective results without causing unreasonable adverse effects on the environment
8. Any specific requirements on reentry to areas where the pesticide has been applied, including in all cases a statement informing the user of the pesticide that the requirements concerning reentry are required to be followed and are incorporated into the label by reference
9. Any specific requirements on the storage and disposal of the pesticide and its

container, including in all cases, a statement that the requirements of EPA are to be followed

10. Any limitations or restrictions on use required to prevent unreasonable adverse effects, such as:

    a. Required intervals between application and harvest of food or feed crops
    b. Rotational crop restrictions
    c. Warnings as required against use on certain crops, animals, objects, or in or adjacent to certain areas
    d. For restricted-use pesticides, indication of which use or uses are restricted to certified applicators or persons under their direct supervision
    e. Other pertinent information which the Administrator determines to be necessary for the protection of health and the environment

Use Classification

All pesticides must bear on the label a statement of the use classification assigned at the time of registration.

General use classification — Pesticides classified for general use shall bear the following statement in the Direction for Use: "General Use Pesticide — Available to the Public".

Restricted-use classification — Pesticides classified for restricted use shall bear the following statement, placed at the top of the front panel, set in type of the same minimum sizes as required for human hazard signal words, and appearing with sufficient prominence relative to other text and graphic material on the front panel to make it unlikely to be overlooked under customary conditions of purchase and use:

1. The statement: "Restricted-Use Pesticide — For purchase and use only be Certified Pesticide Applicators or by persons under their direct supervision"
2. An additional statement when applicable designating the category or categories of certified applicator to whom use is restricted.

Separate labeling — Any pesticide for which some uses are classified for general use and others for restricted use shall be separately labeled according to the labeling standards set forth above and shall be registered and marketed as separate products with different brand names.

PUBLIC CONTROVERSY

In this day of modern science, there is an ever increasing public controversy about the use of herbicides to control weeds. There are increasing demands, justified or not, about the effects of these chemicals on the public health and well-being of people. This fear of the "unknown", perpetuated by certain activist groups, is truly a concern of the weed scientist. He cannot and must not avoid this controversy, but he has a difficult position to defend. He is basically limited to the facts as he sees them and the objective data of his own experiment. Frequently the emotional "problems" of the "opposition" are transferred to the scientist and his work or to the chemical company or to the product.

The public needs and is entitled to greater help in the handling of scientific controversies. It needs and is entitled to assurances as to the credibility and integrity of the process by which decisions are made in scientific disputes involving the public interest. It needs and is entitled to a major by-product of such a process which is assistance in separating out the lay component of the scientific conclusion for lay evaluation by those who choose to make it.[13]

CREDIBILITY IN DISPUTE RESOLUTION

Credibility in dispute resolution, however, requires application of a scientific method just as much to decision *Procedure* — the lawyer's bailiwick — as to ultimate scientific *Substance* — the scientist's bailiwick. The procedural rule of reason is thus in sharp contrast to that part of the traditional adversary legal process that permits and, all too often, actually encourages use of procedural weapons, such as delay, concealment, or personal abuse, for tactical purposes in an effort to reach a desired result.

When a scientist says "the risk is acceptable" (or not acceptable), he is expressing a view made up of scientific observation (data), scientific analysis, scientific opinion — about which there is often little dispute — and personal judgment derived from general background — about which understandably there may be many differences. Usually these two components are confused, subconsciously or unwittingly; sometimes, however, they are deliberately confused by pressure groups for a certain position.

Credibility derives from many things, such as professional qualification and reputation. However, one of its primary ingredients is consistency. Unless properly explained, procedural obstruction may be viewed as inconsistent with asserted confidence in a result determined on the merits. The procedural rule of reason condemns such obstruction. Similarly, even unauthorized or inadequately considered internal confidential statements may be viewed as more expressive of true opinion and attitude than public utterances and positions.

SUMMARY AND CONCLUSIONS

Suppression and/or eradication of unwanted vegetation on or along rights-of-way has been a major phase of chemical weed control. Until recent times, mechanical methods were the only methods available for controlling this vegetation. The use of herbicides in combination with mechanical, cultural, and biological methods of control has greatly expanded the capability to effectively and economically control vegetation in these areas. This has been achieved to a large degree by weed control as a science with herbicides.

REFERENCES

1. **Anderson, W. P.,** *Weed Science: Principles,* West Scientific, St. Paul, Minn., 1976.
2. **Anon.,** *Herbicide Handbook of the Weed Science Society of America,* 3rd ed., 1974.
3. **Anon.,** Selected Weeds of the United States, *U.S. Dep. Agric. Agric. Handbook,* 366, 1970.
4. **Beal, W. J.,** The vitality of seeds, *Bot. Gaz. (Chicago),* 40, 140, 1905.
5. **Crafts, A. S. and Robbins, W. W.,** *Weed Control,* 3rd ed., McGraw-Hill, New York, 1962.
6. **Darlington, H. T. and Steinbauer, G. P.,** The eighty-year period for Dr. Beal's seed viability experiment, *Am. J. Bot.,* 48, 321, 1961.
7. **Fryer, J. D. and Evans, S. A.,** *Weed Control Handbook,* Vol. 1, 5th ed., Blackwell Scientific, Oxford, 1968.
8. **King, L. J.,** *Weeds of the World,* Interscience, New York, 1966.
9. **Klingman, G. C.,** *Weed Control: As A Science,* John Wiley & Sons, New York, 1961.
10. **Muenscher, W. C.,** *Weeds,* 2nd ed., Macmillan, New York, 1962.

11. **Prince, F. S. and Hodgon, A. R.,** Viable Seeds in Old Pasture Soils, Tech. Bull., 89, N. H. Exp. Stn. 1946.
12. **Stevens, O. A.,** The number and weight of seeds produced by weeds, *Am. J. Bot.*, 19, 784, 1932.
13. Pesticide Programs, Title 40, Chapter 1, Subschapter E, U.S. Environmental Protection Agency, Washington, D.C., 1974 (proposed).
14. **Anon.,** Apply Pesticide Correctly, a Guide for Private Applications, U.S. Department of Agriculture and U.S. Environmental Protection Agency, U.S. Government Printing Office, Washington, D.C., 1975.

Chapter 2

THE TECHNOLOGY OF WEED CONTROL*

INTRODUCTION

The safe use of herbicides implies acceptance of an attitude of safety. Safety should be a routine instinctive action rather than an erratic occasional one to meet special requirements. Connotations of safety in terms of a single use are quite complex and are complicated by multiple uses. Certainly the safety of those directly involved, such as the applicator and contracting party, must be considered. More important, though, are the many associated interests which make up the innocent bystander complex.

The following three steps, if taken, will go a long way towards the safe application of herbicides. First, post the area and use all the advance publicity you can obtain to alert the populace of your action. Second, map the area; know the water source, where it goes, and how it is used. Note particularly the direction of prevailing winds. Third, have ready access to constant wind information even to the extent of carrying your own equipment. This is not a complete safety program; it is only meant to stimulate an attitude of safety.

To prevent power interruptions and blackouts, electric utility companies spray herbicides under powerlines for brush and woody plant control. Highway departments spray roadside brush and weeds to improve visibility and road safety; railroads spray herbicides to prevent jamming switches, rotting ties, and cargo damage from overhanging branches. Herbicides keep trails and waterways open and destroy noxious weeds, i.e., poison ivy and poison oak. Forest companies selectively "weed-out" scrub trees in favor of more productive timber and pulpwood species. Ranchers eliminate weeds and brush to open more land for cattle grazing and meat production. The quality and quantity of our food is dependent on chemical weed control.

PLANT TOXICITY STUDIES

The phenoxy herbicides, 2,4-dichlorophenoxy (2,4-D), 2,4,5-trichlorophenoxy (2,4,5-T), MCPA [(4-chloro-o-tolyl)oxy] acetic acid, Silvex®, and related materials, are selective herbicides widely used in the vegetation management of brush trees on rights-of-way.[1] These chemicals are related to naturally occurring plant growth regulators. They kill plants by using malfunctions in growth processes. Broad-leaved plants are generally susceptible to the phenoxy herbicides, whereas most grasses, coniferous trees, and certain legumes are relatively resistant.

These phenoxy herbicides are predominantly toxic to green plants and are much less toxic to mammals, birds, fish, reptiles, shellfish, insects, worms, fungi, and bacteria. When properly used, they do not persist in soils and water at levels harmful to animals and microorganisms. They do not concentrate in food chains and do not accumulate from year to year in croplands. They are detectable only rarely in food and then only in insignificant amounts.

These herbicides are widely used because they are more efficient and usually less hazardous and less injurious to the environment than alternative methods of control. Use of these chemicals is estimated to reduce the cost of production of the crops on which they are used by about 5% and to reduce overall agricultural production costs in the U.S. by about 1%. Uses in forests and nonagricultural situations provide addi-

* By Dr. Philip Phillips, formerly of the U.S. Army Corps of Engineers, North Atlantic Division, New York, New York.

tional savings. If the phenoxy herbicides were no longer available, the cost of food, forest products, electric power, transportation, and governmental services would be higher. These increases in costs would be borne by the consumer.

The phenoxy herbicides were discovered over 30 years ago, incidental to research on naturally occurring plant hormones. As the chemical nature of these substances became known, related compounds were synthesized and tested for effects on plant growth. Certain phenoxy acetic acids, including 2,4-D, were found to be plant growth regulators and in due course were discovered independently in the U.S. and U.K. to be active herbicides capable of selectively killing many broad-leaved weeds in cereal grains, grasslands, and coniferous forests. Further research led to the commercial use of several phenoxy herbicides, including 2,4-D, 2,4,5-T, MCPA, and Silvex®. These pioneer accomplishments have now expanded into a world effort by industrial and public research agencies that have discovered a wide array of useful herbicides, most of which are chemically and functionally unrelated to the phenoxys.

The phenoxy herbicides are weak acids that are only slightly soluble in water and petroleum oils. Although the acid is the active form, they are normally converted into water-soluble amines or oil-soluble esters for convenience in handling and application. These are mixed with other ingredients, such as solvents, emulsifiers, thickeners, and wetting agents, to make commercial formulations for specific uses.

ANIMAL TOXICITY STUDIES

In laboratory tests, greatly exaggerated doses of 2,4,5-T were injected into the stomach and/or fed to pregnant laboratory animals. Hundreds of chemicals and compounds cause harmful effects in pregnant animals when fed in equally high doses: vitamin A, caffein, aspirin, and common table salt; for example, a pregnant woman would need to consume 1/2 cup of pure 2,4,5-T each day for the first 7 months of pregnancy (which would likely be lethal) to approach the laboratory test animal amount. This extreme is impossible under field conditions where spray rates are a very small fraction of those used in the laboratory, and spraying is confined only to target areas.[1]

Before a herbicide can be labeled, it must pass nine lengthy and extensive tests, costing nearly three quarters of a million dollars. These tests include 90-day feeding studies in two different animal species, an inhalation toxicity study in rats, 2-year feeding studies in two rodent species, reproduction studies in two mammalian species, teratology studies in two bird species, acute toxicity in two fish species, mammalian metabolism studies, and mutagenic studies.

A highly poisonous kind of dioxin called tetrachlorodibenzo-*p*-dioxin (TCDD) is an unavoidable contaminant in commercial supplies of 2,4,5-T and Silvex®. The amount present in currently produced formulations of 2,4,5-T and Silvex® is not enough to alter the toxicological properties of these preparations or to endanger human health or to affect plants or animals in the environment.

The presence and significance of this impurity have become known only in recent years. 2,4,5-T formerly contained from 1 to 80 ppm of TCDD, concentrations sufficient to cause skin eruptions in industrial workers called chlorance. Although it has not been feasible to eliminate this contaminant entirely, present production methods are able to reduce the dioxin level to less than 0.5 ppm.[1]

FIELD APPLICATION

To avoid unsightly vegetation, bases of trees may be sprayed without treating the foliage. Brush can be cut, chipped into mulch, and stumps sprayed to prevent resprouting. Brush can be sprayed in winter. Only in difficult terrain where helicopters are used is the temporary brownout unavoidable.

The preferred vegetation on highway and utility rights-of-way is usually grass and other short vegetation. The proper management of vegetation on roadsides is important to safety by providing adequate sight distances on curves and at junctions and intersections. Also, tall weeds can cause hazardous snowdrifts. Before the use of phenoxys, mowing, plowing, hand cutting, and burning were about the only control methods available. 2,4-D, 2,4,5-T, and other herbicides have eliminated the need for plowing, hand cutting, and burning and also permit fewer mowings per year. Chemicals are particularly useful on steep slopes, wet areas, and areas obstructed by guard rails and other structures and not accessible to mowing machines. The use of phenoxy herbicides on rights-of-way is sometimes restricted by the proximity of desirable vegetation susceptible to herbicide injury.

A 20-year study on a Pennsylvania right-of-way shows the population of many common game species to have increased after application: whitetailed deer, turkeys, quail, grouse, rabbits.[2-5] Low-growing food plants and nesting-cover (brackenfern, sweetfern, teaberry, blackberry, and witchhazel) sparked to life after brush was eliminated. Like many other herbicides, 2,4-D and 2,4,5-T are broken down by soil bacteria and disappear in a few weeks.

A diversity of food plants useful to wildlife developed on the right-of-way following spraying. These plants included common herbs of the forest along with invaders such as the common goldenrods and sheep-sorrel. These herbs were highly nutritious and provided food for wildlife both in the summer and winter seasons. Woody plants were interspersed throughout this plant community and included the common blueberries, huckleberry, and teaberry as low shrubs, along with sweetfern and bear oak as tall shrubs or small trees. The taller woody plants supply food throughout the year and are particularly valuable as an emergency food when deep snow covers the ground in the winter. The right-of-way was heavily used by common wildlife species such as whitetailed deer, indicating that attractive food and cover had developed.

AERIAL DRIFT CONTROL

The principal hazard in the use of the phenoxys is to crops and other valuable plants either within the treated area or nearby. Treated crops and forest trees can be injured through accidental overdosing, improper timing of treatments, unusual weather conditions, and other causes. Injury to nearby crops and ornamentals can result from drift of droplets or vapors of the spray. Such losses are largely preventable through the use of proper formulations and spray equipment and the exercise of good judgment. The Microfoil Boom®, the Directa-Spray®, and Lo Drift® Spray Additive (Amchem Products, Inc., Ambler, Pa.) can control and reduce drift to a minimum.[6-8]

These attempts to achieve better control over the placement of herbicidal sprays have also resulted in the use of various low-pressure spray systems, special flooding nozzles, and other devices designed to produce a spray pattern characterized by a greater amount of large droplets. In addition, more careful consideration of meteorological conditions at the time of spraying has succeeded in reducing damage due to spray drift.

While these practices have led to improvement in spraying techniques, there is still a definite need for more effective methods of spray placement in areas with mixed livestock production and cropping systems. In such situations, hormone-type herbicides can be applied only during that season of the year when there is little danger of damage to crop plants.

Recently, another effort to reduce the hazard of drift has been attempted through alteration of the principal physical characteristics of the spray liquid. This has been accomplished primarily by the use of highly viscous, crowded-phase, herbicide emulsions which resist break-up into fine droplets and promote larger droplet formation.

A coined, nondescriptive term, "invert emulsion", has been applied to such viscous sprays and implies a phase inversion from the more common oil-in-water emulsions to the water-in-oil types wherein the oil assumes the continuous phase and water assumes the dispersed phase.

In a continuing effort to increase herbicide safety, Amchem devised the Microfoil Boom® (Plates 1 to 3).* This is an aerial spray boom designed to mimimize air turbulence at the point where spray droplets form, thus controlling their size mechanically. With the microfoil, droplets of nearly uniform and predictable size can be produced from standard water, oil, or oil-in-water emulsions.

The Microfoil Boom® is constructed in 3- and 5-ft sections; 6-in.-wide airfoil-shaped nozzles are mounted on the boom. Each nozzle contains 60 needle-like orifices along its trailing edge. The following orifice sizes are available: 0.013-in. ID which produces droplets with a volume mean diameter (VMD) of 800 μm; 0.028-in. ID which produces droplets with a VMD of 1700 μm; and 0.060-in. ID which produces droplets with a VMD of 4000 μm. Variations from these means is approximately ±100 μm. With 120 orifices per foot of boom, in flight, the microfoil produces a sheet-like spray pattern. This spray falls to the ground before it can be disturbed by rotor blast.

Performance reports consistently confirm that under normal spraying conditions, microfoil droplet size and virtual elimination of fine spray particles restrict herbicide contact to the intended treatment area. Drift injury to adjacent or distant plants is essentially eliminated. Brush control from microfoil treatments has been excellent.

Microfoil Booms® are utilized commercially for herbicide applications to utility rights-of-way, drainage ditches, pastures and rangelands and for forest site preparation, plantation establishment, and conifer release purposes for which their relatively large droplets and coarse drift distribution are suitable.

A study was made to compare drift from a Microfoil Boom® and a conventional boom, both helicopter mounted.[9] An UV-sensitive dye was used to determine spray deposits. Cards were placed at 10-ft intervals on a 100-ft-high tower 200 ft from the flight line and at regular intervals on the ground. Each boom was flown over the line and at an altitude of 50 ft. Wind velocity was 10 mi/hr. Conventional boom application left droplets on the cards for the full height of the tower, thus indicating the presence of particles 100 ft or more in the air at least 200 ft from the flight line. After the microfoil pass, no droplets could be found on the tower cards.

STABLE AIR DRIFT STUDIES

A test was planned to measure downwind deposits from microfoil sprays applied in air inversions. This was conducted near Woodland, Calif. as part of extensive airborne drift control studies.[10,11] Red dye emulsion was sprayed from a helicopter-mounted microfoil over a flight line in a rice field. Multiple runs were made at an altitude of 10 ft to simulate applying a herbicide to rice. Mylar sheets were placed downwind from the treatment swath to collect dye spray deposits, and various air-sampling devices were operated during and after spraying to collect any of the particles suspended in the air for several miles downwind. Spray deposits were photoelectrically analyzed to determine dye content as a measure of particle drift.

Applying the spray from a helicopter-mounted 0.013-in. orifice Microfoil Boom® essentially eliminated drift-prone droplets smaller than 900 μm. A total of 98 to 99%

* All plates appear following page 32.

of the spray material was recovered in the treated field. The downwind deposit corrected for an initial application of 1 lb/acre was less than 1 μg/ft² (equal to 0.043 g/acre) 100 ft from the flight line.

A greenhouse study to check the effect at these levels of 2,4-D butoxyethanol ester on several crops known to be sensitive to the compound was made.[12] Cotton, tomato, tobacco, soybean, cucumber, and snapbean plants were treated at rates of 10, 1.0, 0.1, and 0.0 μg/ft², equal to 0.43, 0.043, 0.0043, and 0.0 g of 2,4-D acid equivalent per acre.

Epinasty 24 hr after exposure has been established as a reliable indicator of 2,4-D drift interception. Examination of the treated plants 24 and 72 hr after 2,4-D treatment showed slight (less than 30°) epinasty of the third trifoliolate petiole of snapbeans at the 10-μg rate, but no detectable injury to other species at any of the rates used. Examination 7 days after application showed no additional epinasty, nor was there any evidence of characteristic 2,4-D symptoms in the new growth.

These results indicate that even during strong air inversions, drift from this application of 2,4-D ester at 1 lb/acre would not be detectable on sensitive plants 100 ft or more from the spray swath.

CROSSWIND DRIFT STUDY

A cooperative study designed to measure the extent of downwind drift from greater flying heights during moderate to severe crosswinds was conducted near McAllen, Tex. A 2,4-D/picloram solution containing 6 lb of 2,4-D acid equivalent and 1.62 lb of picloram acid equivalent as triisopropanolamine salts in 12 gal of water per acre was applied from a Bell G-4 helicopter equipped with a 0.028-in. orifice Microfoil Boom®. The flight line was 1000 ft long; swath width was 50 ft. Two flights each were made at altitudes of 70 and 120 ft, the heights normally used for spraying distribution and transmission lines, respectively. At these heights, the pilot measured crosswinds of 15 to 20 mi/hr. Winds at the anemometer 12 ft from the ground were 4 to 8 mi/hr during the runs.

Bioassay stations were established 100, 200, 400, and 800 ft from the flight path, in a line parallel to the wind direction. Six potted "Black Valentine" snapbean seedlings were placed at each station and were returned to a greenhouse after the day's runs. In 24 hr after spraying, the number of bean plants showing epinasty at each station was recorded.

Snapbean plants at the 100- and 200-ft stations showed epinasty characteristic of 2,4-D when examined 24 hr after spray application. No epinasty was evident at the stations 400 and 800 ft from the flight line, except a trace at the 800-ft station from one of the 120-ft altitude applications (see Table 1).

The results indicate that under these test conditions of crosswinds severe enough to halt normal spray operations, injury to sensitive plants could be expected up to 200 ft from the intended treatment area, but beyond that, even at high wind velocities, drift injury would decline sharply.

DRIFT CONTROL WITH ROTARY DISKS

The unique design of the Directa-Spray® disk (Plates 4 to 6) combines simplicity of operation with good drift control. Conventional water-soluble sprays and special spray materials, such as invert of oil-water emulsions and wettable powders, may be applied. It has many advantages over conventional spray booms. It is light in weight; when you hold the applicator in your hands, you are holding the equivalent of a 25- to 50-ft boom. The mechanism is simple to install; one clamp or one bolt will fasten

Table 1
NUMBER OF SNAPBEAN PLANTS SHOWING EPINASTY 24 HR AFTER POTENTIAL EXPOSURE TO 2,4-D[a]

Altitude (ft)	Feet from flight line 100	200	400	800
70	5	3	0	0
70	5	2	0	0
120	3/3	5/5	0	Trace
120	3/3	0/3	0	0

[a] Six plants per station unless otherwise noted; "2/3" indicates that two of three plants at the station showed epinasty.

From Akesson, N. B., Wilce, S. E., and Yates, W. E., *Agrichem. Age,* 13, 10, 1979. With permission.

Table 2
CALIBRATIONS[a]

RPM	Swath width (ft) 25	30	35	40	50
46	21	22	23	24	25
58	20	21	22.5	24	27
70	21	22	23	24	27

[a] Nozzle for feet ground canted up at 10° angle spraying water. Note, under ideal wind conditions, angle may be increased to 20° to obtain a 20% in swath width.

From Akesson, N. B., Wilce, S. E., and Yates, W. E., *Agrichem. Age,* 14(12), 11, 1971. With permission.

the applicator to the vehicle from which you wish to spray. One man can operate the applicator system from two switches in the truck cab, one to turn on the spray unit and the other to turn on the spray solution. Nozzle clogging is eliminated. You can apply any formulation with the applicator: wettable, emulsifiable, or soluble. The applicator can be operated from any spray equipment delivering at least 15 gal/min at 40 lb/in.^{2} Table 2 shows swath widths that can be expected as a function of pressure and revolutions per minute (rpm). Pressure has much greater effect on swath than rpm. Pressure narrows or widens the swath, as does holding the unit above or below horizontal. Invert emulsions or thickened materials will increase swath width. Table 3 shows gallons per minute as a function of pressure and rpm, based on water only. Actual formulation used will cause some variation.

Table 3
GALLONS PER MINUTE

	Output (gal/min)				
RPM	25	30	35	40	50
46	10	10.5	11.5	12	13
58	9.5	10.5	11.5	12	13
70	9.5	10	11	11.5	—

From Akesson, N. B., Wilce, S. E., and Yates, W. E., *Agrichem. Age,* 14(12), 11, 1971. With permission.

DRIFT CONTROL WITH SPRAY ADDITIVES

Lo-Drift® spray additive (Plates 8 and 9) is a new product. When added to conventional herbicide formulations, the droplet size is increased, thereby reducing the hazard of drift.[8] It is neither a foam nor a particulating agent. Chemically, it is a water-soluble polymer. As such, it has the ability to unite the tiny particles that make up the spray solution to which it is added. This ability substantially reduces the number of fine spray particles produced while spraying. Changing the physical characteristic of the spray mixture also enables the spray material to adhere more effectively to the foliage upon which it is sprayed.

Spray mixtures can be applied with most conventional aerial and ground spray equipment. No special nozzles or equipment are required. Mixing directions on the label should be followed carefully. First mix the herbicide in a spray tank half filled with water. Agitate thoroughly for 1 min prior to adding the recommended amount to the spray mix and continue to agitate the spray mix for 5 to 10 min prior to adding the remainder of water to the spray tank. There are a number of similar drift control agents which may be added to a herbicide spray solution to control drift.[12]

DRIFT CONTROL BY DROPLET SIZE SELECTION

Damage claims blamed on 2,4-D injury during the past 30 years are substantial. In spite of the outlawing of 2,4-D in many places, the use of 2,4-D has grown in both pounds of acid and acres treated.[1,13-15] The study of applying 2,4-D with less hazard of drift has been emphasized. Work with frilling tree injectors, granular formulations, thickening agents, and invert emulsions has led to the decrease in hazard of these materials to nearby desirable plants.

To show how invert emulsions may be sprayed on target, the drift contained, and the pesticides on nontarget areas minimized, the following must be taken into account:

1. Stokes' law and how its equation may be used as a fundamental principle in predicting drift
2. Requirements of coverage to kill plants
3. Techniques and equipment used with the bifluid system to meet these requirements

The drift of a droplet is the distance a crosswind will blow it during the time required for the droplet to fall. The equation for Stokes' law in this form is

$$V = \frac{2gr^2 (p - p_1)}{9n}$$

where v = velocity at which a particle drops, g = force of gravity, r = radius of the particles, p = density of the particles, p_1 = density of air, and n = viscosity of air.

By using a combination of units from each set and making a couple of assumptions; one can reduce this equation to a very simple and usable form:

$$D = \frac{1.49 \times 10^4 VH}{R^2}$$

where D = drift distance (ft), V = wind velocity (mi/hr), H = drop distance (ft), and R = droplet diameter (um).

One can calculate how far a droplet of 100 μm (which is typical of droplets in a conventional spray system) will drift if it is sprayed at a height of 50 ft with a crosswind of 1 mi/hr. The equation then becomes:

$$D = \frac{1.49 \times 10^4 \times 50 \times 1}{(10^2)^2} = 75 \text{ ft of drift}$$

One can make a similar calculation for a droplet of 590 μm, which is the mass median diameter produced by the Rhap-Trol® (Rhodia, Inc., Chipman Division, New Brunswick, N.J.) spray system. The equation becomes:

$$D = \frac{1.49 \times 10^4 \times 50 \times 1}{(5.9 \times 10^2)^2} = 2 \text{ ft of drift}$$

To consider conditions which may be encountered when spraying a right-of-way with a helicopter, the distance that droplets of 100, 300, 590, and 800 μm will drift when sprayed from a height of 50 ft in crosswinds of 2, 5, and 10 mi/hr are given in Table 4. The 300- to 800-μm droplets are in a comparatively safe range, even in a 10-mi/hr wind, for all are contained within 83 ft. To consider the same particles sprayed at a height of 5 ft which approximates roadside spraying, the predictable drift is given in Table 5. From these data, it is seen that a size of 300- to 800-μm approaches a higher degree of safety.

The above calculations seem to imply the desirability of having droplets of 800 to 100 μm in size. For the sake of controlling drift, this would be excellent. However, for practical application, the most important factor is droplet spacing.

As droplet diameters are halved, the number of droplets increases eightfold. Table 6 shows the number of droplets 1 gal would theoretically cover per square inch if uniformly sprayed over an acre, assuming each droplet to be the size stated. This table also gives the expected drift when sprayed at a height of 50 ft in crosswinds of 10 mi/hr. Droplets of 300 to 800 μm can be sprayed from a helicopter equipped with Rhap-Trol® spray system. This spray will give good kill at a low application rate; e.g., 4 to 12 gal of product gives an adequate number of droplets per square inch as shown in Table 6. Thus, droplets large enough to contain the drift, but still small enough to spray the required number of droplets per square inch to kill, are now a commercial reality. Visko-Rhap® (Rhodia, Inc., Chipman Division, New Brunswick, N.J.) herbicides are used in this spray program.

Table 4
PREDICTABLE DRIFT OF DIFFERENT PARTICLE SIZES WHEN SPRAYED AT 50 FT

Diameter of particles[a] (μm)	Speed of crosswinds (mi/hr) 1	5	10
100	75 ft	275 ft	750 ft
300	8 ft, 4 in.	42 ft	83 ft
590	2 ft, 2 in.	10 ft, 8 in.	21 ft, 5 in.
800	1 ft, 3 in.	5 ft, 9 in.	12 ft

[a] 1 μm = 1/25,000 in. or 1/1000 mm.

From Hill, L. O., Fundamentals of Drift Control by Droplet Size, Tech. Rep. 9, Appendix D, U.S. Army Engineer Waterways Experiment Station, Vicksburg, Miss., 1975.

Table 5
PREDICTABLE DRIFT OF DIFFERENT PARTICLE SIZES WHEN SPRAYED AT 5 FT

Diameter of particles (μm)	Speed of crosswinds (mi/hr) 1	5	10
100	7 ft, 6 in.	37 ft, 6 in.	75 ft
300	9 in.	4 ft, 3 in.	8 ft, 4 in.
590	3 in.	1 ft, 1 in.	2 ft, 2 in.
800	2 in.	2 in.	1 ft, 3 in.

[a] 1 μm = 1/25,000 in. or 1/1000 mm.

From Hill, L. O., Fundamentals of Drift Control by Droplet Size, Tech. Rep. 9, Appendix D, U.S. Army Engineer Waterways Experiment Station, Vicksburg, Miss., 1975.

Table 6
COMPARISON OF NUMBER OF DROPLETS AND DRIFT FOR VARIOUS PARTICLE SIZES

Diameter particles[a] (μm)	Calculated number of droplets per square inch per gallon per acre	Calculated drift — 50-ft height in 10-mi/hr wind
100	1164	750 ft
300	52	83 ft
590	5.6	21 ft, 5 in.
800	2.6	12 ft
1000	1.1	7 ft, 6 in.

[a] 1 μm = 1/25,000 in. or 1/1000 mm.

From Hill, L. O., Fundamentals of Drift Control by Droplet Size, Tech. Rep. 9, Appendix D, U.S. Army Engineer Waterways Experiment Station, Vicksburg, Miss., 1975.

DRIFT CONTROL BY INVERT EMULSIONS

In many areas, irrigation, drainage, navigation channels, and various types of water reservoirs are severely infested with noxious vegetation. Shoreline brush and tree control may also be a problem. Commercial use of the Stull Bifluid System as an application medium has achieved outstanding weed control as reported in trade publications technical conferences, etc. While most weeds and brush are susceptible to various herbicide formulations, effective control is often necessary immediately adjacent to crops or ornamentals which are equally sensitive to the hormone herbicides.[14,15]

Success of the Stull Bifluid System and of BIVERT-PH® and BIVERT-HCE® (Stull Chemical Co., San Antonio, Tex.) as adjuvants with phenoxy herbicides for the control of emergent weeds is due to several factors. First, the active ingredient is surrounded, or encased, in a microthin oil film which does not disperse readily in water and consequently reduces overall water contamination. Second, the spray mixture is a viscous, water-in-oil (invert) emulsion which affords low-drift characteristics. Third, the invert emulsion has the unique quality of floating on the water surface, thus, spray droplets which do not impact directly on vegetation float in the water surrounding the weeds until they contact plant stems or leaves and are absorbed. In these ways, the maximum amount of active herbicide is concentrated on the target weeds where it does the most good.[17]

To summarize other studies, only 1% of the spray volume from tests of invert emulsions is composed of droplets smaller than 200 μm. In comparison, standard emulsion patterns averaged 29.7% spray volume of droplets smaller than 200 μm.

Droplet size and quantitative determinations made by means of deposit collection, spray interception, and air sampling techniques of fine particle characteristics of standard emulsion sprays as contrasted to the coarse droplet patterns of invert emulsions are illustrated in Figure 1.

Deposit samples showed average maximum mean droplet size to be 198 μm for invert emulsions and 244 μm for standard spray. Downwind samples of mean droplet size beyond 35 to 50 ft showed a range of 166 to 333 μm for inverts and 100 to 299 μm for conventional emulsions. The average maximum droplet size was 6933 μm for ester invert emulsion, 6483 μm for the amine invert, and 1033 μm for the standard spray.

Steeply sloping deposit curves were characteristic for all functional runs conducted with invert emulsions. Deposit levels fell below 1.0 mm^2 of relative volume (0.01 gal/acre) at distances of 20, 35 to 50, and 50 to 80 ft downwind in wind velocities of 4, 4 to 6, and 6 to 10 mi/hr, respectively. Deposit volumes for all runs of conventional oil-in-water emulsions showed a large positive margin beyond 50 ft downwind when compared to deposit patterns of invert emulsions.

More than 99% of invert emulsion spray volume was deposited no farther than 100 ft downwind for all fully functional runs as compared to 82.4% for standard spray runs.

Interception samples were very effective in characterizing the horizontal movement of spray from the treated area. Large droplet fractions of the spray pattern rapidly settled out as deposit, and with wind velocities of less than 10 mi/hr, no droplets larger than 500 μm were detected beyond 70 ft from the application line. Conventional emulsion runs were responsible for interception of numbers of droplets ranging from 456 to 640 μm at given stations, while numbers from invert emulsions gave volumetric interception curves which fell below the 1.0-mm^2 level within limits of the sample line, while an average of 72.5 mm^2 of relative volume of standard emulsion was still airborne at 500 ft from the application line. At 400 ft, 380 times more standard spray than invert emulsion was present in the air for runs conducted during comparable wind velocities.

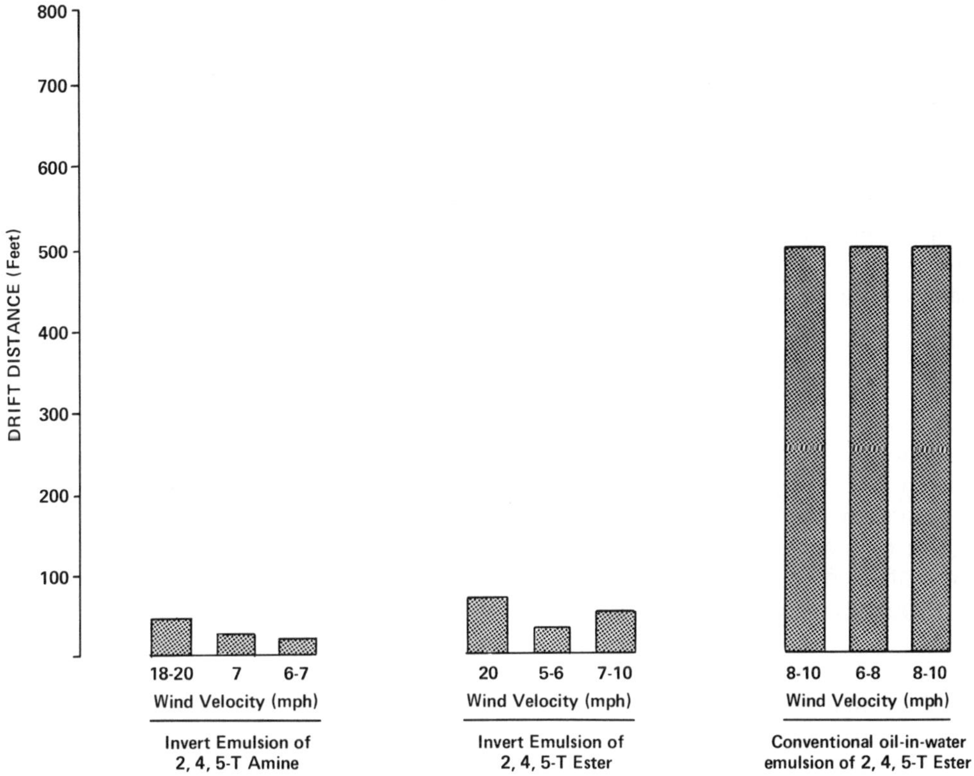

FIGURE 1. Differential herbicidal drift under varying wind conditions as shown by bioassay.[17]

Air-sampling procedures resulted in detection of abundant numbers of fine-particle (25 to 125 μm) standard spray at all air-sampling stations. However, droplet size values for invert emulsions were in the 200- to 500-μm size classes, with extremely low sampling frequency at collection stations. No functional run with invert emulsion left traces of spray, 700 ft downwind, 30 ft in height, but standard runs resulted in an average relative volume of 77.1 mm^2.

Tomato plants were used as a bioassay of herbicidal drift. A general relationship was developed wherein the deposit level of 1.0 mm^2 of relative volume was established as the marginal tolerance level of deposit with respect to herbicidal damage to tomato plants. From measurements of epinasty and terminal phytotoxicity, it was determined that in wind velocities of 18 to 20 mi/hr, invert emulsions of 2,4,5-T ester drifted 48 ft downwind from the edge of the effective spray swath. Runs conducted with invert emulsions in operational wind velocities (less than 10 mi/hr) resulted in a damaging drift to 33 ft downwind from the edge of the effective swath. Damaging amounts of conventional oil-in-water emulsions of 2,4,5-T ester drifted 470 ft downwind, the maximum distance for placement of test plants.

Abundant spray traces noted on the air sampler (700 ft downwind) and herbicidal damage noted on ornamental plants 1000 ft downwind indicated that damaging drift from conventional spray was present to a distance of at least 1000 ft downwind.

From the results of these studies, comparing the performance of high viscosity emulsions vs. standard oil-in-water sprays, it is evident that the bifluid spray system and the chemical formulations used therein are extremely effective in reducing the quantity and extent of herbicidal spray drift (see Figure 1).

No consistent differences were determined between the performance of ester and amine forms of 2,4,5-T applied as invert emulsions relative to the physical characteristics of the spray pattern or the developmental growth patterns of test plants.

It is further concluded that standard oil-in-water emulsions of 2,4,5-T applied with coarse atomization are hazardous to use in close proximity to susceptible plants, even with low-velocity wind conditions, and at distances of 1000 ft or more.

The large droplet size characteristics of invert emulsions applied through bifluid equipment and the resultant positive deposit pattern attainable makes possible an accurate pesticide placement heretofore unattainable through the use of conventional, aqueous-phase spray mediums.

BIFLUID SYSTEM MECHANICS

Chemical Dose Rate

An application of 2 lb of acid equivalent per acre is sufficient for any single application on emergent aquatic weeds susceptible to the phenoxy herbicides applied through the Stull Bifluid System. Individual weed species differ in their susceptibility; therefore, local use practices should dictate which specific compound is selected. Low-volatile ester emulsifiable concentrates are recommended, since they are not water soluble and can be concentrated better on emergent species.

Application-Mixing Procedure

The Stull Bifluid System requires two fluid holding tanks. One contains an oil-phase premix, and one contains a water-phase premix. The oil-phase premix consists on one of the application adjuvants, BIVERT-PH® or BIVERT-HCE®, and Number 2 diesel oil. The water-phase premix contains the phenoxy herbicide emulsifiable concentrate in water. These premixes are automatically proportioned and mixed as they are used through a small device (Stull Bi-Vac unit) located in the suction side of the sprayer pump. To reduce tank requirements on boat equipment, two Bi-Vac units may be used. One unit will proportion and mix the emulsifiable concentrate from the shipping container with water directly from the stream or reservoir and then inject this water phase premix into the second Bi-Vac located in the pump suction.

Application Volume Rates

Two recommended application rates are given, *minimum* and *maximum*. The quantities of materials given under each column are in even gallons for convenience of visualizing the relative proportions of BIVERT®, diesel, phenoxy herbicide concentrate, and water which are automatically proportioned and dispensed by the sprayer. In every case, the sprayer will be applying 2 lb of phenoxy acid equivalent per surface acre, if a 4-lb equivalent phenoxy ester concentrate is used (see Table 7).

Equipment

Any sprayer modified to bifluid application standards may be employed, including aircraft, power sprayers, and wind turbines. For all types of equipment, spray tips should be selected which produce minimum droplet size consistent with the desired draft control.

Coverage

Good coverage is essential to good weed control. The application volume rates recommended in Table 7 will provide adequate volume to obtain optimum kills. The choice of minimum vs. maximum gallonage should be determined by the degree of drift control desired. Use minimum rates with small droplets where drift is not a critical

Table 7
APPLICATION VOLUME RATES

Ingredients and rate	Relative gallon mixing ratios: Minimum rate	Maximum rate
Aircraft or Fixed Boom		
Oil-phase premix		
BIVERT-HCE®	1	
BIVERT-PH®		1
Number 2 diesel oil	2	5
Make:use rate	3:2	6:3
Water-phase premix		
Phenoxy herbicide emulsifiable concentrate	1	1
Water	27	41
Make:use rate	28:14	42:21
Total water-in-oil emulsion per acre volume rates	16	24
Broadcast or Handgun		
Oil-phase premix		
BIVERT-PH®	1	1
Number 2 diesel oil	9	7
Make:use rate	10:5	8:15
Water-phase premix		
Phenoxy herbicide emulsifiable concentrate	1	1
Water	89	269
Make:use rate	90:45	270:135
Total water-in-oil emulsion per acre volume rate	50	150

From Peterson, D. P., Hays, O. H., and True, C. R., New Techniques in Vegetative Maintenance on Military Reservations, Tech. Rep. 9, Appendix E, U.S. Army Engineer Waterways Experiment Station, Vicksburg, Miss., 1975.

Table 8
SELECTION AND INSTALLATION OF METERING ORIFICES

Sprayer output (gal/min)	Minimum size water suction components (in.)	Minimum size oil suction components (in.)	Water orifice diameter (in.)	Oil orifice diameter in inches for indicated water-oil ratio: 6:1	9:1
1.00—1.75	3/4	1/2	0.140	0.059	0.046
1.80—3.00	3/4	1/2	0.180	0.075	0.061
3.10—5.00	3/4	1/2	0.250	0.103	0.083
5.10—10.00	3/4	1/2	0.344	0.140	0.115
10.10—18.00	1	1/2	0.469	0.182	0.151

Note: For example, to determine water- and oil-phase metering orifices required for sprayer output rate of 4.8 gal/min at a water-to-oil phase ratio of 9:1, (1) locate sprayer output rate (4.8 gal/min) in left column of chart and (2) move horizontally to right and find 0.250 water-phase orifice and 0.083 oil-phase orifice. The water-to-oil ratios listed are approximate only. Variations on the order of 10% are considered normal and acceptable. With a given set of metering orifices, do not attempt to apply the spray emulsion at a discharge rate outside the range shown for those orifices.

From Peterson, D. P., Hays, O. H., and True, C. R., New Techniques in Vegetative Maintenance on Military Reservations, Tech. Rep. 9, Appendix E, U.S. Army Engineer Waterways Experiment Station, Vicksburg, Miss., 1975.

factor. Use maximum rates with large droplets where the drift factor is critical. Selection and installation of metering orifices is given in Table 8.

REFERENCES

1. Council for Agricultural Science and Technology, The phenoxy herbicides, *Weed Sci.*, 23, 253, 1975.
2. **Bramble, W. C., Byrnes, W. R., and Worley, D. P.,** Effects of certain common brush control techniques and materials on game food and cover on a power line right-of-way No. 3, *Pa. Agric. Exp. Stn. Prog. Rep.*, 151, 1956.
3. **Bramble, W. C. and Byrnes, W. R.,** Ecological aspects of brush control — a long-term study on a utility right-of-way, *Purdue Univ. Agric. Exp. Stn. Res. Bull.*, 835, 1967.
4. **Bramble, W. C. and Byrnes, W. R.,** Fifteen years of ecological research on a utility right-of-way, in Proc. N.E. Weed Control Conf., Cornell Ornamentals Research Laboratory, Farmingdale, N.Y., 1969, 270.
5. **Bramble, W. C. and Byrnes, W. R.,** Long-term ecological study of game food and cover on a sprayed utility right-of-way, *Purdue Univ. Agric. Exp. Stn. Res. Bull.*, 885, 1972.
6. Amchem Production, Inc., Aerial Drift Control with the Microfoil Boom, Technical Service Data Sheet, Agricultural Chemicals Division, Ambler, Pa., 1972.
7. Amchem Products, Inc., Directa-Spray disk, Technical Service Data Sheet, Agricultural Chemicals Division, Ambler, Pa., 1974.
8. Amchem Products, Inc., Lo-Drift Spray Additive, Technical Service Data Sheet, Agricultural Chemicals Division, Ambler, Pa., 1974.
9. **Kirch, J. H., Waldrum, J. E., and Bishop, P. W.,** The Microfoil, an aerial device for controlling drift from conventional sprays, *Proc. South. Weed Control Conf.*, 22, 385, 1969.
10. **Akesson, N. B., Yates, W. E., and Wilce, S. E.,** Controlling spray atomization, *Agrichem. Age,* 13(12), 10, 12, 16, 1970.
11. **Akesson, N. B., Wilce, S. E., and Yates, W. E.,** Confining aerial applications to treated fields — a realistic goal, *Agrichem. Age,* 14(12), 11, 13, 1971.
12. **Kilingman, G. C. and Ashton, F. M.,** *Weed Science; Principles and Practices,* Interscience, New York, 1975.
13. **Olenik, C. J.,** The wonderful power of selectivity to power line rights-of-way, *Weeds Trees Turf,* 16, 14, 1977.
14. U.S. Environmental Protection Agency, Rights-of-Way Pest Control, U.S. Government Printing Office, Washington, D.C., 1976.
15. **Gangstad, E. O., Ed.,** Integrated Control of Alligator Weed and Water Hyacinth in Texas, Tech. Rep. 9, U.S. Army Engineer Waterways Experiment Station, Vicksburg, Miss., 1975.
16. Rhodia, Visko-Rhap herbicides, Technical Report, Chipman Chemical Division, New Brunswick, N.J., 1975.
17. **Phillips, P. J.,** An Evaluation of High Viscosity Crowded Phase Emulsions, as Herbicide Carriers when Applied Through the Bifluid Spray System, Project Ap-7b, Southwest Agricultural Institute, San Antonio, Tex., 1963.
18. **Hill, L. O.,** Fundamentals of Drift Control by Droplet Size, Tech. Rep. 9, Appendix D, U.S. Army Engineer Waterways Experiment Station, Vicksburg, Miss., 1975.
19. **Peterson, D. P., Hayes, O. H., and True, C. R.,** New Techniques in Vegetative Maintenance on Military Reservations, Tech. Rep. 9, Appendix E, U.S. Army Engineer Waterways Experiment Station, Vicksburg, Miss., 1975.

Chapter 3

WEED CONTROL AUTHORITIES AND REGULATION

INTRODUCTION

Although we all have a general concept as to what a weed is, there are times when we are hard put to determine whether a particular plant is a weed or not. Most plants become weeds when they grow in places where they are specifically not wanted for a justified reason. On the other hand, a number of plants commonly thought of as weeds may actually serve some useful purpose, such as soil erosion control, wildlife food and cover, or a niche in a plant succession.

Authority for control and eradication of undesirable species of plants have been made a part of law, and a number of public agencies are involved in this effort. Palliative measures are available and, in most cases, are not prohibitive in cost if current technology and reasonable discretion in methods of application are followed.

SCOPE OF WEED CONTROL PROBLEMS

Weeds are a menace to the industrial and agricultural economy of our nation and a threat to the health of humans, livestock, fish, and wildlife. They compete with desirable economic plants for water, light, mineral nutrients, and other plant-growth requirements. Some weeds harbor insect vectors of human diseases, serve as a reservoir for disease-producing organisms, and are poisonous to humans and livestock.

No one knows how many "weed species" populate the earth, but an estimated 30,000 species are known to be widely distributed throughout the world. Most of our major cultivated crops are subject to infestation by as many as 200 weed species. Between 20 and 30 species cause serious problems and must be controlled each year[2,5,6] (see Appendix A for common and Latin names, and botanical authority).

The scope of weed-control operations on privately owned lands in the U.S. is enormous. Cultural, mechanical, ecological, and biological methods of weed control are used on more than 400 million acres of cultivated crops each year. The intertilled row crops receive two to five seedbed tillage operations and an average of about five cultivations each year. More than 1 billion acres of hay, pasture, and rangelands, and several million acres of nonagricultural lands, aquatic sites, and rights-of-way benefit by weed-control practices.

Federal and state agencies are responsible for the management of about 1 billion acres of land, all of it closely associated with privately owned land that is used for crop production, grazing, and forestry. Most of this 1 billion acres is in various stages of weed infestation. More than 10 million acres of this land are infested by weeds classified as noxious by the state and federal seed laws. In addition, millions of acres in road, railroad, and utility line rights-of-way, fence rows, ditchbanks, waterways, and industrial sites are infested with weeds.

Unless weeds on federally and state managed lands, rights-of-way, industrial sites, and other nonagricultural areas are controlled, it will be difficult to reduce losses caused by weeds on privately owned lands. Such public lands provide an instant source of reinfestation. This annual reinfestation maintains populations of weed seed, increases annual losses, requires increased use of herbicides, and increases the cost of control by all methods.

Poison ivy, poison oak, and other poisonous plants are a menace to many unsuspecting persons seeking recreation. Each year poisonous weeds cause nearly 2 million cases of skin poisoning and other skin irritations, all of which adds up to a loss of 333,000 working days per year. Poisonous weeds also cause 3.7 million days of restricted activity and 1/2 million days spent in bed. The cost of medical care caused by ragweed and other pollens to which hundreds of thousands of people are allergic is very extensive. The misery caused by poison ivy, poison oak, and toxic pollens is equally serious[1,3,4,7,8]

FEDERAL AND STATE LEGISLATIVE AUTHORITY

The following federal and state laws and regulations affect the importation, domestic movement, or quarantine of noxious weeds:

1. Federal Plant Pest Act (administered by the U.S. Department of Agriculture [USDA]) — prohibits the importation and interstate movement of parasitic plants and seeds (including dodder, mistletoe, witchweed, etc.)
2. Interstate shipment of animals, birds, fish, and plants (administered by Department of Justice) — prohibits the interstate movement of waterhyacinth, alligator weed, and water chestnut (including seeds)
3. Federal Seed Act (administered by USDA) — supplements the state seed laws in interstate commerce; prohibits the occurrence of noxious weed seed in excess of the requirements of the state into which seed is shipped, whether from foreign or domestic sources
4. Federal Seed Act Regulations (administered by USDA) — prohibits the importation of white top, Canada thistle, dodder, quackgrass, Johnsongrass, bindweed, Russian knapweed, perxenmial sowthistle, and leafy spurge in agricultural and vegetable seed
5. State seed laws (administered by individual states) — all states have seed laws designating certain weed seeds as noxious weeds

RULE MAKING REGULATIONS

A ready reference to federal legislative authority (United States Code [USC], Code of Federal Regulations [CFR], and Supplemental Authority) pertinent to the importation, domestic movement, quarantine, control or eradication of noxious weeds and their propagating parts is given below:

I. Title 7 — Agriculture
   - A. Chapter 3. Federal Plant Pest Act of 1957 prohibits the importation or interstate movement of parasitic plants (administered by USDA).
   - B. Chapter 6. Provides for the control of competitive shrubs to permit growth of adequate desirable vegetative cover for soil protection on range or pasture land (administered by USDA).
   - C. Chapter 7. Provides for the control of noxious weeds in order to protect soil or water resources (administered by USDA).
   - D. Chapter 37. Seeds
     1. Prohibits interstate shipment of certain seeds that do not meet specific labeling requirements.
     2. Prohibits importation of certain seeds if adulterated or unfit for seeding purposes.
   - E. Chapter 40. Halogeton Glomeratus Act authorizes the Departments of Agriculture and the Interior to take all necessary steps in the control, suppression, and eradication of this weed (administered by USDA and USDI).

PLATE 1. Microfoil Boom® with a 0.028 nozzle.

PLATE 2. Closeup of the 0.028 nozzle.

PLATE 3. Droplet formation with the 0.028 nozzle.

PLATE 4. Directa-Spray® nozzle, applying brush killer to dormant brush on a utility rights-of-way.

PLATE 5. Closeup of Directa Spray® nozzle.

PLATE 6. Closeup of Directa Spray® nozzle in use.

PLATE 7. Conventional nozzles in use for conifer release.

PLATE 8. Conventional nozzles in use with Lo Drift® added to the spray solution to reduce drift hazzard.

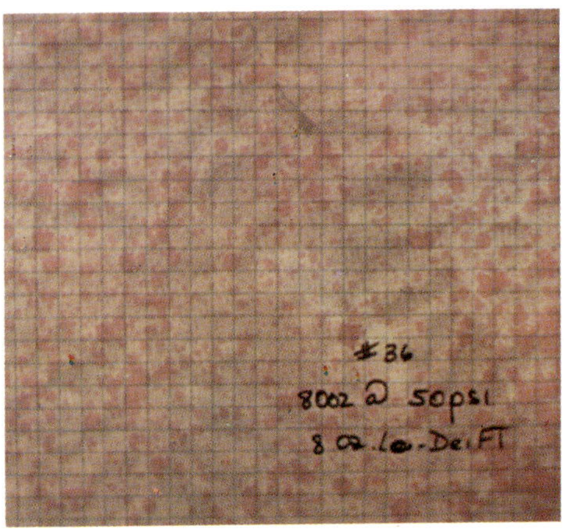

PLATE 9. Spray pattern with Lo Drift® added to the spray solution.

PLATE 10. Backpack mistblower.

PLATE 11. Tree injector.

PLATE 12. Stump or basal spray operation.

(Photography courtesy of Roy Johnson, A. Chemical, Inc., Ambler, Pa.)

F. Chapter 3. Federal Noxious Weed Act of 1974 (Public Law 93-629) provides for the control and eradication of plants of foreign origin, new or not widely prevalent in the U.S., and the regulation of movement in interstate or foreign commerce.

II. Title 16
Tennessee Valley Authority Act of 1933 — weed control operations are carried out under authority of this act (administered by Tennessee Valley Authority [TVA]).

III. Title 18
Chapter 3. Crimes — Animals, birds, fish, and plants — places the interstate movement of waterhyacinth and related plants in the crime category (administered by Department of Justice).

IV. Title 33
Chapter 9. Protection of Navigable Waters — provides for the control and progressive eradication of waterhyacinth, alligatorweed, Eurasian watermilfoil, and other obnoxious aquatic plant growths (administered by Corps of Engineers).

V. Title 40
Protection of Environment (Subchapter E, Pesticide Programs) — regulations for enforcement of the Federal Insecticide, Fungicide, and Rodenticide Act (FIFRA), Public Law 92-516 (86 Stat. 973) and amendments (1975 and 1978).

VI. Title 50
Chapter 1. Wildlife and Fisheries — prohibits the introduction of plant life into any wildlife refuge area without approval of appropriate federal authorities (administered by USDI) and provides authority for control of undesirable plants.

History of the Development of Laws and Regulations

In the U.S., we have had laws affecting pesticide development, sale, and use dating back to the original Federal Food and Drug Act (see Table 1). Subsequent laws were passed which resulted in stricter, more comprehensive and more specific regulations. The most comprehensive of all is the Federal Environmental Pesticide Control Act which was signed into law on October 21, 1972. The major provisions are as follows:

1. Will be administered by the Environmental Protection Agency (EPA)
2. Requires all U.S. pesticides to be federally registered or approved
3. Classifies pesticides for general or restricted use, the latter, because of hazard to the individual or the environment; to be applied only by or under the supervision of certified applicators
4. Establishes State Applicator Certification programs (with farmers certified as private applicators) and cooperative enforcement programs
5. Prohibits the misuse of pesticides whether or not personal or crop injury results or residue exceeding the tolerance is found at harvest, adds civil to increase criminal penalties, otherwise strengthens enforcement
6. Shortens administrative review procedures
7. Requires pesticide producing establishments to be registered and to regularly submit information on production and sales volumes

Table 1
HISTORY OF LAWS AND ACTIONS AFFECTING THE DEVELOPMENT OF AGRICULTURAL CHEMICALS

June 30, 1906 —	Original Federal Food and Drug Act. This act was intended to prevent the manufacture, sale, or transportation of adulterated, misbranded, poisonous, or deleterious foods, drugs, medicines, and liquors (administered by the FDA of the USDA).
April 26, 1910 —	Original Federal Insecticide Act. An act to prevent the manufacture, sale or transportation of adulterated or misbranded insecticides and fungicides (administered by USDA).
June 25, 1938 —	The revised Federal Food, Drug, and Cosmetic Act. Enacted to prohibit the movement in interstate commerce of adulterated and misbranded food, drugs, devices, and cosmetics; this law charged the FDA with the responsibility of keeping poisonous insecticides, as well as insects, out of food products. It included provisions for setting tolerances. The law did not provide for advance clearance of safety of food additives. It left it up to the FDA to discover their use, and to make tests to prove them ''poisonous or deleterious'' to the satisfaction of the court when action was required to remove them from the market.
1945 and 1946 —	Many new insecticides (organic) appeared on the market. Legal tolerances were not announced, indicating to many that the ponderous tolerance procedures of the Food, Drug, and Cosmetic Act could not handle the flood of new materials. In addition, the labeling requirements of the 1910 Insecticide Act were obsolete.
June 25, 1947 —	The Federal Insecticide, Fungicide and Rodenticide Act (FIFRA). This act extended the scope of the 1910 Act to include rodenticides and herbicides and required for the first time premarketing registration of products to be shipped interstate. It thus set up the system of label claims and registration. It provided that a complete copy of the label and a statement of all claims to be made for the material must be submitted to the Secretary of Agriculture and must be approved before the material is registered for sale (in interstate commerce). It also provided a system of fees to support the procedures involved. The act was enforced by the USDA, ARS, Plant Pest Control Branch, and the Pesticide Regulation Section.
January 1950 —	The continued flood of new material emphasized the need for residue tolerances. According to the provisions of the 1938 law, supporting evidence for the tolerances had to be established at public hearings. Accordingly, hearings were started on January 17, 1950 and were completed in September 15, 1950. Such a mountain of evidence was presented that several more months were required to digest it.
January 1951 —	Growing public concern over the addition of chemicals to food products led to the appointment of the Delaney Committee, known formally as the House Select Committee to Investigate the Use of Chemicals in Foods and Cosmetics.
June 1952 —	The Delaney Committee submitted its report to the 2nd session of the 82nd Congress. The report was in three parts: (1) a majority report, indicating that present laws were inadequate and new laws were needed, (2) a minority report, indicating that present laws were adequate but needed streamlining, and (3) a supplementary report indicating that one of the most important hazards involved was in the use of pesticides in the home garden on materials intended for home consumption. It was pointed out that this particular hazard is beyond the reach of law and is susceptible only to education of the general public.
November 1952 —	The Food Protection Committee of the National Research Council published a comprehensive report, dealing with the use of chemicals in food. This report considered the present and future food needs of the nation and stressed the vital role that chemicals play in meeting these needs. It pointed out that the losses from hazards involved in using chemicals are rather minor compared to the losses involved in not using them.

Table 1 (continued)
HISTORY OF LAWS AND ACTIONS AFFECTING THE DEVELOPMENT OF AGRICULTURAL CHEMICALS

March 26, 1953 —	Representative A. L. Miller introduced a bill, H.R. 4277, which provided for the regulation of pesticides and tolerances. This bill was heralded by the industry as a basis upon which a workable law could be built.
January 11, 1954 —	Representative Miller submitted a new bill, H.R. 7125 (a revised H.R. 4277 of the previous session). It was referred to the House Committee on Interstate and Foreign Commerce and became known as the Miller Pesticide Residue Amendment, since it was introduced as an amendment to the Federal Food, Drug, and Cosmetic Act.
July 22, 1954 —	The Pesticide Chemicals Amendment (The Miller Pesticide Residue Amendment) amended the Federal Food, Drug, and Cosmetic Act to provide a system of fees to support the procedures involved and became Public Law 518 of the 83rd Congress. It became effective for new chemicals a year from the date of signing and was effective for tolerances established as a result of the 1950 and subsequent hearings. It established new and more expeditious procedures for obtaining tolerances for exemptions for pesticide chemicals. Though administered by HEW, it specifically assigned to the Secretary of Agriculture the determination of usefulness of a chemical for all, any, or none of the uses for which tolerances or an exemption are sought. It required the USDA to express an opinion as to whether proposed tolerances are reasonable and whether residues are likely to result from proposed patterns of use.
September 6, 1958 —	Food Additives Amendment to the Federal Food, Drug, and Cosmetic Act (Public Law 929 of the 85th Congress). It established procedures, principles, and appeal procedures for obtaining regulations from HEW prescribing safe conditions of use for food additives. A food additive was defined as any substance the intended use of which may reasonably be expected to result directly or indirectly in its becoming a component of or otherwise changing the characteristics of any food. The law specifically excluded pesticide chemicals, pesticide residues on crops (raw agricultural commodities), and pesticide residues in processed food when such residues result from legal uses of pesticides on crops. These pesticides are regulated under Public Law 518 (the Miller Amendment). This food additives amendment included the *Delaney Clause on carcinogens:* no additive shall be deemed safe if it is found to *induce cancer* when ingested by *man or animals.*
March 27, 1962 —	Declaration of Certain Forms of Plant and Animal Life Viruses to be Pests. Thus, it extended the scope of the Federal Insecticide, Fungicide, and Rodenticide Act to cover some kinds of pesticides not previously subject to the Act.
May 12, 1964 —	An Amendment to FIFRA eliminating registration under protest. This amendment provided that the registrant could request an advisory committee or public hearing when registration of his product was refused, canceled, or suspended. The Secretary of Agriculture was also given the authority to require, by regulation, the USDA registration number of the label.
October 21, 1972 —	The Federal Environmental Pesticide Control Act (FEPCA) of 1972. It significantly amended the FIFRA of 1964 and is administered by the EPA.
November 28, 1975 —	The Federal Insectide, Fungicide, and Rodenticide Act of 1975. It significantly amended FIFRA of 1972 and is administered by the EPA.
September 30, 1978 —	The Federal Insecticide, Fungicide, and Rodenticide Act (FIFRA) as amended, established new regulations for registration and reregistration of pesticides.

8. Authorizes indemnification of certain owners of pesticides which are suspended, and canceled
9. Authorizes the EPA Administrator to establish pesticide packaging standards, regulate pesticide and container disposal, issue experimental use permits, conduct research on pesticides and alternatives, and monitor pesticide use and presence in the environment

Pesticide Registration Procedures

The law requires the registration of pesticides for two primary purposes:

1. To insure the product will be effective for the purposes claimed on the label when used in accordance with the directions
2. To ensure that the product will pose no safety hazards to humans or unreasonable adverse effect to the environment when label directions and precautions are followed

The various forms and data which are required by the EPA of an applicant are discussed in detail in "Guidelines for Registering Pesticides in the United States". Copies of the latest printing of this document are available from Pesticides Regulation Division, Office of Pesticides Programs, Environmental Protection Agency, Washington, D.C. 20250.

Four basic types of data areas are normally required in obtaining a clearance so that a tolerance may be established. They are

1. Performance data or field testing, including information on the pesticide formulation, pests, crop, and the exact use
2. Residue data, including pesticide residue levels for the exact use and metabolite studies with an analytical method suitable for enforcement
3. Environmental data, including residue toxicology information on wildlife and possible pollution in soil, water, and air
4. Toxicology data, including acute and chronic toxicity studies on rats, dogs, and other organisms depending on the use of the control agent

A person or manufacturer applies for registration of a pesticide under FIFRA by submitting to EPA a petition proposing establishment of tolerance levels for a particular pesticide use based upon data in the four areas listed above.

The tolerance is based on the maximum residue expected from the proposed use. A tolerance level must be a safe level, i.e., there must be a practical certainty that residues on the food within the tolerance level will impose no health hazard.

Prior to December 2, 1970, the Food and Drug Administration, under the authority of the Federal Food, Drug and Cosmetic Act, was responsible for the establishment of tolerances (or exemptions from tolerances) for pesticide residues in human food or animal feed intended for interstate shipment. While EPA now establishes the tolerance, as prescribed by FIFRA, FDA is still responsible for enforcing tolerances.

Petitions are requests for the establishment or exemption (or revocation) of tolerances. Tolerances are established after review of petitions. Most petitions are submitted by pesticide manufactures, but they also come from formulators, government agencies, user groups, or other interested parties. Each petition must be accompanied by a fee as prescribed by the regulations. However, fees are usually waived for petitions submitted by universities or government agencies upon request. The types of data required in a petition are spelled out in the guidelines.

Once a tolerance has been established, application for registration may be submitted for additional uses of the pesticide on the commodity provided they do not leave residues in excess of the safe tolerance level.

Types of Registrations

Registration of new herbicides may be for two types of products:

1. "New chemicals" are products comprised of, in whole or in part, chemicals not previously registered for any purpose. The level of review effort for a new chemical product calls for maximum of 45 to 60 man-days of scientific review, including safety, efficiency, and usefulness of the product and the overall impact of the use of the product on the ecology. Generally, multiple contacts with the company are required to assemble and clarify data. This can have considerable influence on total review time.
2. The second classification of new products is one not previously registered for the particular company submitting the application, although similar or identical products have already been registered for other companies.

Except where a new use is involved, products in this category need considerably less scientific review time than new chemicals because of the already documented safety and efficacy data assembled on similar or identical products already registered by other companies.

Renewal Registrations

Regulations issued under FIFRA require registration renewal every 5 years. Under these regulations, 3,200 products became eligible for renewal in fiscal year (FY) 1971, and 9,000 renewals were due in FY 1972 from an original eligible 11,200. This activity was suspended during early 1972 due to restricted pesticides. These renewals are expected to be processed at the time of classification and reregistration of all products under the new legislation. An expected registration lapse of 20% is expected each year. Applications for renewal are subject to any new or revised standards since the product was registered and have not been applied to the registration during the period. Although the review may not always be as extensive as for a new product, it would include information not available with a new product such as use experience over the preceding 5 years, including accident and incident reports and information on environmental safety. Follow-up investigational samples of these marketed pesticides for chemical and biological analyses which have been almost negligible in the past will be increased as pesticide field investigators are established in regional offices.

AMENDMENTS TO EXISTING REGISTRATIONS

Amendments to existing registrations:

1. Added or deleted uses
2. Minor changes in formulation
3. Changes in directions for use on a label
4. Modifications in warnings and cautions in labeling
5. Other modifications in labeling, i.e., company name, product name, etc.

Temporary Tolerancc

A temporary tolerance or temporary exemption may be requested. These are usually granted for a period of 1 year to run concurrently with a temporary permit to cover experimental use of the pesticides on a specified number of acres in specified geographic locations. The temporary permit allows the petitioner to test the pesticide on a relatively large scale without destroying the crop. It also allows the treated crop to be marketed legally. Since consumer exposure will be quite limited, the toxicity and residue data required for granting of a temporary tolerance are usually less than for a permanent tolerance.

Clearance and Labeling

Once a tolerance has been set, a pesticide is considered cleared for use on the particular crop against a certain pest as indicated in the application for registration. However, before the chemical can be marketed and used, it must be labeled. FIFRA requires that pesticide products be properly labeled prior to being introduced into interstate commerce. Now, with the amendments contained in FIFRA, products sold in intrastate commerce must have a federal label as well.

Labeling is generally defined as all labels and other printed or graphic matter attached to or accompanying the pesticide product or to which reference is made on the label or accompanying literature. The requirements to be met for label approval are detailed in the "Guidelines for Registering Pesticides in the United States".

Public Announcements

Under FIFRA as amended, the pesticide user should have more registration information available and options to act upon than in the past. The Act requires that the EPA Administrator publish several kinds of notices, starting with the registration or denial of registration of a particular pesticide and including items such as any change in classification and any proposed cancellation. The notices, by and large, are required by the Act to be published in the *Federal Register.*

CONTROL AND ERADICATION OF WEEDS

Weed control by federal agencies is confined for the most part to work on federally managed lands and allied waterways. Weed control usually comes under the broad authority of each agency for land management and improvement. Purchases of herbicides and the cost of weed control usually come out of federal appropriations for land and water management programs.[9]

The Corps of Engineers has specific authority for certain aquatic control operation. The National Agricultural Conservation Program, as approved by the Secretary of Agriculture (pursuant to the authority vested in him under Sections 7 to 17 of the Soil Conservation and Domestic Allotment Act, as amended), is authorized to carry out certain types of weed and brush control, under the Soil Conservation Program, on a cost-sharing basis with farmers.

The Agricultural Research Service (ARS) and the USDI are jointly authorized to administer the Halogeton Glomeratus Control Act.[10] The Rural Electrification Administration (REA) and the USDA do not own or maintain any electrification or communication facilities. However, REA administers loan programs which enable individual borrowers to develop and build their own systems under REAs management and technical assistance. The application of herbicides is a part of the rights-of-way maintenance, but is paid for by the borrowers.

The Departments of Agriculture, Interior, Health, Education, and Welfare, Army, Air Force, Navy, Post Office, and agencies, such as the Veterans Administration, Atomic Energy Commission, Tennessee Valley Authority, General Service Administration, and National Aeronautics and Space Administration, conduct weed-control operations on public lands as part of their land management programs.

A survey of state legislative authority for weed control indicates that some type of noxious weed-control regulations have been in effect for many years. The Pennsylvania statutes were first enacted in 1862, and Illinois recognized the need for weed control in 1872, and Wisconsin recognized the need in 1887. A law to tax individuals for noxious weed control has been in effect in Idaho for more than 40 years, in Indiana since 1929, and in Nebraska since 1936. In a number of states, the revision and updating of weed-control laws are underway, stimulated by the rapid development of more effective herbicides and methods of application. Noxious weed-control laws of one kind or another are in effect in 25 states.

These laws, however, vary widely both in scope and effectiveness. In most of the 25 states, the declared noxious weeds are determined by laws or regulations. However, in some states, county commissioners make the determination. The number of noxious weeds varies from 1 species in some states to as many as 70 in other states. In most states, the list of weeds declared noxious under the seed laws differs considerably from those so designated by the noxious weed statutes; the latter are generally fewer in number. A number of states combine the control of noxious weeds with other pests, such as insects and plant and animal diseases. In most states, noxious weed laws are administered by the Commissioner (or Secretary or Board) of Agriculture or his staff. His authority, however, varies widely in different states. In some he has advisory functions, in others his approval of county or district programs and methods of control is required, and in still others his powers are more sweeping and binding. In a few states, a college or university or a department thereof is an integral part of the noxious weed-control program. In a much larger number of states, such participation, if any, is advisory only. To a larger degree, the extension service at both the state and county levels has part in noxious weed control, generally on an advisory committee basis. However, in at least one state, the extension service, by law, has a leading part in administering the control program, both at the state and county level. In a number of states, the law requires that farm machines, such as harvesters, hay balers, etc. — especially those operated on a custom basis — shall be swept "reasonably free" of weeds and litter before being moved upon the highway. A few states similarly regulate the movement of manure, fertilizers, weeds, and seeds to prevent the movement of noxious weeds through these channels.

RESEARCH ON WEEDS AND THEIR CONTROL

The USDA has a continuing long-term program of basic and applied studies directed toward the solution of weed problems on farms, forests, rangelands, irrigation canals, ponds, rights-of-way, and other areas where weeds and brush are problems. Research on reducing damage caused by weeds includes studies to develop biological, chemical, cultural, ecological, mechanical, and integrated methods of control. Research is also conducted on life histories and growth patterns of individual weeds. Weed research includes physiological and biochemical studies to determine the mechanisms involved in absorption and translocation, the mode and site of action, and the effect of environment on plant responses to herbicides. Comprehensive studies are made to develop principles, practices, and methods of using herbicides and other weed-control techniques in solving regional agricultural weed and brush problems in agronomic crops, horticultural crops, forests, grazing land, and aquatic and noncropland sites. All of this research is conducted cooperatively with state agricultural experiment stations, with the exception of the research at Beltsville, Md., Denver, Colo., Fort Lauderdale, Fla., and Mayaguez, P.R. There is also cooperation with federal agencies such as the Bureau of Reclamation and Bureau of Land Management, Department of the Interior and the U.S. Army Corps of Engineers.

Cooperation ranges from informal exchange of information to formal arrangements providing funds and personnel in support of weed control research. Industrial companies cooperate in furnishing experimental chemicals, equipment, and funds essential to rapid progress in weed-control investigations. Certain private, semipublic, or public groups furnish financial support for cooperative research on weed problems.[14]

EDUCATIONAL PROGRAMS ON WEEDS AND THEIR CONTROL

People are the key to preventive weed control. The Federal-State Cooperative Extension Service is the link between research and regulatory agencies and the people who are responsible for the control of weeds. Four million farmers and ranchers, 33 million home owners, personnel in federal, state, and local government agencies, and pesticide dealers and applicators make up the composite of people responsible for the control of weeds.

The general public needs to understand the importance of weeds — their effects on the national economy and the health and welfare of the people of the U.S. Responsible personnel must be familiar with the technology and regulations concerning weeds and herbicides. They also need to be taught the skills and mechanics of weed control; the precision required in the use of herbicides and the precautions essential to safe use must be widely understood.

It will be difficult, if not impossible, to lower weed populations and reduce annual weed-control costs significantly unless the effectiveness of methods of control is improved through regulatory, research, and education programs.

We must improve the effectiveness of methods of control so that the excessive use of chemical and/or nonchemical methods of control can be avoided. If this were accomplished, we could prevent the physical impairment of the soil by excessive cultivations that increase wind and water erosion and herbicide contamination of soil and water.

It is recommended that (1) all federal agencies develop improved procedures for analyzing, recording, and reporting activities related to weeds and their control, and (2) research, education, and training programs be expanded by appropriate federal agencies to support improved regulatory and action programs on weed control.

THE FEDERAL PESTICIDE ACT OF 1978

On September 30, 1978, President Carter signed into law the Federal Pesticide Act of 1978, amending the Federal Insecticide, Fungicide, and Rodenticide Act (FIFRA). That was the final step in a process which had begun some 18 months previously with the submission of Administration proposals to the Congress. Congress and the EPA, which has chief responsibility for administering FIFRA, have worked together to refine this legislation.

The amendments build and improve upon all the major areas covered by the previous law.

1. Registration (premarket clearance) of pesticides to prevent unreasonable hazards to humans or the environment
2. Classification of pesticides for general or restricted use
3. Certification of users of restricted-use pesticides
4. Informative and accurate labeling of pesticide products
5. Enforcement of FIFRA to ensure proper pesticide practices

The cornerstone of the new, more efficient approach authorized in the amended law is a program of generic pesticide registration. In the past, registration entailed an examination of risk of each product one at a time. However, generic registration of pesticides entails a single, comprehensive evaluation of risks and benefits of the technical material common to numerous products based on all data relevant to the registration decision. The new system will demand less time and money both from EPA and the registrant.

Development of pesticide generic standards will take time, and until such standards are developed and the complete new reregistration program is put in place, Congress has authorized a program for EPA to grant conditional registrations. Conditional registration will allow EPA to process applications for registration of new products which are like ones already registered and thus permit the new products to enter the market on an equal footing with others already registered and in turn provide consumers with a wider range of comparable products. Ultimately, all products will be reviewed comprehensively when reregistered under generic pesticide registration standards.

New uses of "old" chemicals and new chemicals will also be eligible for conditional registration if EPA determines that enough information is available to evaluate unique hazards that may be posed by the new uses. The Agency is also authorized to issue conditional registration of new chemicals if the public interest would be served by a registration and if risks during the period required to complete and submit additional studies are not unreasonable. Conditional registration will be useful for allowing time to meet new data requirements or in some special cases, for permitting early registration of new chemicals.

DATA REQUIREMENTS

In addition to authorizing generic and conditional registration, the new law clarifies many matters relating to use of data:

1. Compensation by one registrant for use of data owned by another registrant
2. Exclusive use of data submitted for registration of new pesticide chemical
3. Several issues pertaining to efficacy data, and registration data for "minor uses"

All data submitted from 1970 on are eligible for compensation for a period of 15 years following the submission of the data. This means that subsequent applicants must offer to pay compensation to data originators in order to rely on their data. If registrants cannot agree on fair compensation, the dispute will be settled by binding arbitration. Previously the Agency was responsible for settling disputes over compensation.

The new law will also give 10 years "exclusive use" of data to the original registrant after a new chemical is first registered. This means that no one can use the data for 10 years without express permission of the data developer. This aspect of the law rewards innovation in the industry and may be especially useful in encouraging development of biological pesticides and other compounds for which patent protection was difficult to achieve in the past.

The amended law provides that the basic health and safety data underlying pesticide registrations are open to the public, and thus the public will have access to the information which EPAs regulatory decisions are based. Other data, however, are now specifically protected from public disclosure. EPA or other government workers, including those under contract to the government, who wrongfully disclose such data are subject to imprisonment or a fine.

Data as to the efficacy of a pesticide have generally been required for registration. EPA now has relatively broad discretion to waive submission. This new policy is more or less an experiment in permitting the marketplace to police efficacy. If it does not stand the test of time, EPA will reconsider a new approach, as appropriate. In any case, the agency will continue to consider efficacy when performance of the product bears upon public health.

The new law also gives states broader authority and responsibility for registering pesticides. States automatically have authority to register products for use within a state for "special local needs". Formerly, states had that authority only with EPA approval. EPA may disapprove state registration if the use is dissimilar from federal registration, if an imminent health hazard exists, or if the state has authorized use on crops for which no tolerance or allowable residue level on crops has been established by EPA.

Congress has amended two other separate but related areas in FIFRA — classification of pesticides and applicator certification.

In order to make full use of the state programs for training and certification of applicators to use "restricted-use" pesticides, EPA is authorized to classify uses by regulation, apart from the registration process. This will help realize the objectives of the applicator certification program, since almost all states have worked hard to train and certify applicators to handle the restricted products. Restriction of use also gives EPA an option short of cancellation to reduce pesticide risks. Also in connection with the training and certification programs, EPA is authorized to pay up to 50% of the costs for each state program.

In a legal opinion from the office of the Chief of Council of the Chief of Engineers, applicators of pesticides who are federally certified are not required to obtain a state permit, even though the program is delegated to states, but must conform to the particular requirements of the state program.

For users of all pesticides, the new amendments have clarified an issue of broad concern: the legality of pesticide uses or practices not addressed in the label directions. Several important pesticide use practices have been specifically excluded from the definition of "use inconsistent with the label", and the Agency may identify other uses which will be considered consistent though not in literal accord with the label.

Pesticide exporters must now comply with special label requirements including a requirement for pesticides not registered for use in the U.S. Importers must also sign a statement acknowledging that they know such products are not registered in the U.S., a copy of which must be sent to an appropriate official in the importing country.

REFERENCES

1. ACS, Cleaning Our Environment; the Chemical Basis for Action, *American Chemical Society,* Washington, D.C., 1969, 249.
2. **Cain, S. A.,** *Biotope and Habitat of Future Environments,* Natural History Press, Garden City, N.Y., 1966.
3. Council for Agricultural Sciences and Technology, The phenoxy herbicides, *Weed Sci.,* 23, 259, 1975.
4. **Crafts, A. A.,** Herbicide behavior in the plant, in *The Physiology and Biochemistry of Herbicides,* Audus, L. J., Ed., Academic Press, New York, 1964.
5. **Danielson, L. L., Ennis, W. B., Jr., Klingman, D. L., Shaw, W. C., and Timmons, F. L.,** A Survey of Extent and Cost of Weed Control and Specific Weed Problems, ARS 34-23-1, Agricultural Research Service and Federal Extension Service, U.S. Department of Agriculture, Washington, D.C., 1965, 23.
6. **Garman, W. H.,** Agriculture's place in the environment; considerations for decision making, *J. Environ. Qual.,* 2, 327, 1973.
7. National Research Council, *Pest Control Strategies for the Future,* National Academy of Sciences, Washington, D.C., 1972, 375.

8. **Sheets, T. J. and Danielson, L. L.**, Herbicides in soils, in The Nature and Fate of Chemicals Applied to Soils, Plants and Animals, ARS 20-9, Agricultural Research Service, U.S. Department of Agriculture, U.S. Government Printing Office, Washington, D.C., 1960, 170.
9. River and Harbor Act, Sections 104, 1965, 33 USC 610, October 27, 1965, U.S. Army Corps of Engineers.
10. Halogeton Glomeratus Control Act, Title 7, United States Code, Sections 1671 to 1656, July 14, 1952, U.S. Department of Agriculture, Washington, D.C.
11. Public Law 92-516, 21 October 1972.
12. Public Law 94-140, 28 November 1975.
13. Public Law 95-396, 30 September 1978.
14. Report of the USDA Research Planning Conference on Aquatic Weed Control, Davis, Calif., September 1977.

Chapter 4

FEDERAL REGISTRATION AND REREGISTRATION OF PESTICIDES*

INTRODUCTION

There are three important points to begin this discussion

1. Why were pesticide registration regulations developed?
2. What are the economic effects of these regulations?
3. What are some of the new aspects and guidelines of the new pesticide law?

First we will discuss some general points about why our federal registration system was developed. As thinking individuals, we know that to live in a society there have to be some types of laws governing its members. Industry is not against such laws. In fact, industry is for them because without law there would be anarchy with no society and no industry. Members of industry, as other segments of the population, also inhabit this spaceship earth, breathing the same air, eating the same food, and enjoying lakes, streams, and ponds as well as nature in general.

However, industry is against unnecessary laws and regulations. This is the crux of the matter. Government agencies interpret the law either to operate more efficiently or to obtain the authority for some course of action they deem desirable; in so doing they establish rules or regulations by publication in the *Federal Register*. Practically speaking, industry must operate under these rules and regulations that have the force of laws. When government agencies go beyond the original law, industry is against such over-regulation. The thesis is that we now have enough laws and regulations governing pesticides to make their development and marketing as safe as is practical. We seem to have a national syndrome that, when a problem arises affecting many people, someone says, "There ought to be a law". Problems are not solved by laws or regulations, but by people who perceive the problems and then work to obtain a solution.[1-3]

POPULAR PRESS INTERPRETATIONS

Fertilizers, fungicides, insecticides, and herbicides are important reasons for our modern technology in agriculture. Then came Rachel Carson's *Silent Spring*. This moving, well-written book focused attention upon possible ecological problems and aroused genuine concern. This book, however, was better written than researched. Several of the author's contentions, while written with great sincerity and conviction, were simply not accurate. Shortly after this, agricultural chemicals and particularly insecticides and herbicides came under a tremendous attack from a vociferous minority, often uninformed or at best, but partially informed. Half-truths, untruths, errors of fact, and unscientific extrapolations from a preliminary experiment or limited and ambiguous data are accepted as "gospel truth" and quoted and requoted in the popular press.

Our modern urbanized, big-city society has little agricultural knowledge or intimate experience in agriculture. In the absence of knowledge, rumors and fears have fertile ground in which to grow. We are currently living in a period in which many members of the general public are afraid about their food supply. Our society has forgotten that

* By Wendell R. Mullison (consultant), Dow Chemical Company, Midland, Michigan.

it was only a short time ago that spices were very important in world trade, not so much for enhancing the flavor of food, but to cover up the taste of decaying meat. Instead of realizing that we in this country have the most varied, cleanest, safest, and the highest-quality food in the world, many people have turned to various types of food fadism, often accompanied with antichemical and antipesticide views. This antipesticide attitude has had unfortunate side effects apart from food production.

Scientists working in or associated with food production know that there is no difference between protein synthesized by a plant using ammonium sulfate as a nitrogen source and protein synthesized by a plant using manure as a nitrogen source. Yet stores or departments that handle "natural foods" are growing in numbers and popularity even though the price is two or more times the cost of food produced by modern agricultural means, utilizing agricultural chemicals in its production.

It is our firm conviction that all of us, whether it be academic, government, business, or production, have a duty and obligation to educate the general public about modern agriculture. We must arm the general public with real knowledge to the facts. Ecology is a very involved subject, cutting across many disciplines. While reporters strive for accuracy, they cannot be technically competent in all fields, particularly one as complex as ecology. We strongly urge that all of us take an active part in telling the public about modern agriculture. There is a great agricultural story that can be told. Then people will not be misled by premature release of scanty and imcomplete experiments or scare headlines and stories which may be designed to sell newsparers and magazines rather than to inform.

Toxicity concept of toxicity is the first point. Everything is toxic at some concentration. Paracelsus in the early part of the 16th century phrased it very succinctly in Latin which in a free translation may be stated, "Toxicity is a function of concentration". There are many excellent examples. Salt is essential to the human diet, but it has caused death. A few years ago a tragic accident resulted in a New York hospital when several babies died because salt had been substituted for sugar in the babies' formula. Salt has also been used as a method of committing suicide by the Chinese. This is the same concept (but in reverse) in drug or chemical dosage — response curves which states that there is a dosage at which a chemical has no apparent effect.

Carcinogeneses

A second point is that we must defuse the current scare words of carcinogenic, oncogenic, mutagenic, and teratogenic. There is evidence today that shows many chemicals administered in high enough dosages and under proper experimental conditions can cause cancer. It is quite likely that this will also be true for teratogenic effects. For example, vitamin A has been reported to cause teratogenic effects. However, these dosages and experimental conditions are so far removed from the conditions to which you and I would be exposed that they can be safely ignored.

Zero Tolerance

As a third point we must bring out the problem with the concept of zero tolerance. The analytical skills of our chemists are so sophisticated these days that they can detect very, very minute amounts of chemicals. This is particularly true in the measurement of residues. It is nearly always possible to find a residue if enough time is spent to develop a sophisticated analytical method. The interpretation, however, of the significance of such small quantities is another question. One part per million is 1 min in 2 years (1.901 years), and 1 part per billion is 1 min in 1900 years. To make this more meaningful, it is 1 min since the birth of Christ. Now it is even sometimes possible to analyze in parts per trillion. However, measuring such small quantities requires extremely careful procedures and sophisticated techniques to obtain reliable results.

There is a real problem in being certain that what is measured is what it is supposed to be, not an interfering substance. In any case, the magnitude of such quantities is difficult to appreciate, and in the majority of cases, residues in the parts per billion and smaller ranges are only of academic interest.

Zero Risk

Finally, there is the concept that you cannot prove a negative. There is always some unanswered question or a set of conditions that can be postulated for which data are not available. The manufacturer cannot prove the product will be 100% safe under all circumstances. The best that can be hoped for is to prove that the product works and is reasonably safe to use when the label instructions are followed. When driving a car we do not insist on absolute safety.

Congress is responsive to the general public, and the reason we now have new laws regulating pesticides is because the general public was afraid and wanted the old law changed.[4] If we allay the fears and concerns of the public regarding food and agricultural chemicals, and educate people to know that we have the most wholesome and nutritious food in man's history, and that we have pesticides that are not only effective but safe to use, Congress will respond. However, to accomplish this, all of us must take every opportunity to educate the public and thus give it the knowledge that is power. When this is done, registration regulations will reflect the changed attitude, and regulatory agencies will be able to listen to the rule of reason rather that political rhetoric or misguided statements.

ECONOMIC CONSIDERATIONS

Speaking of registration regulations brings up the second part of this paper, i.e., some economic data which may be of special interest.[4,5] First, let us state that the time now necessary to bring a new agricultural chemical from the time of discovery to market is 8 to 10 years. The costs of doing so have risen markedly as shown in Table 1. Table 2 shows the rate of development of major pesticides in the world from 1931 to date. Table 3 shows the changes in the way our research dollar has been spent in recent years. Table 4 shows the length of time it has taken us to obtain tolerances for pesticides, registration, and changes in registrations.

These tables show that there is a decrease in the number of pesticides being introduced. There is a smaller proportion of money available for discovering a new pesticide or to find new uses for old ones. The time necessary to obtain approval is very long. Finally, these regulations do one more thing that is rarely mentioned: they force pesticide discovery and development to be in the hands of large corporations. This is against the innovative small entrepreneurs. Society suffers lack of progress and increased food costs because big corporations do not have all the brains or ideas for developing new products.

The new government regulations and guidelines have a direct bearing on cooperative work with government agencies. This brings up the third and final part of this paper: namely, federal requirements for registration regarding efficacy although the proof is industrial producer. The Environmental Protection Agency (EPA) requires the following information in reports to demonstrate the efficacy of pesticides:

1. Crops (varieties), pest, and/or site treated
2. Location of the tests
3. Plot size
4. Number of replications
5. Soil type and organic matter

Table 1
COST OF DEVELOPING AN AGRICULTURAL CHEMICAL

Year	Millions of dollars	Chance for a new chemical to become a product
1956	1.2	1/1,800
1964	2.9	1/3,600
1969	4.1	1/5,040
1971	10.0	1/7,000
1975	13.6	1/15,000

Table 2
MAJOR PESTICIDES INTRODUCED IN THE WORLD FROM 1931 TO DATE

Decade	Number of pesticides introduced
1931—1940	1
1941—1950	9
1951—1960	18
1961—1970	19
1971—Date	3

Table 3
RECENT TREND IN PERCENTAGE OF RESEARCH DOLLARS FOR DISCOVERING AND DEVELOPING PESTICIDES AT DOW

Year	Discovery research (%)	Development and registration (%)
1970	53	47
1971	48	52
1972	43	57
1973	42	58
1974	39	61
1975	38	62
1976 (est.)	34	66

Table 4
AVERAGE TIME NECESSARY FOR DOW TO OBTAIN REGISTRATION OR TOLERANCES IN THE U.S. 1970-1975

Type of regulatory action	Number of applications	Time in weeks
Minor label or formulation change	175	18
New relatively minor use	61	51
Obtaining tolerances and registration of a new product or a major new use	40	131

6. Formulation used
7. Rate of application
8. Finished spray volume per acre
9. Time of application and in terms of crop stage
10. Stage of growth of weeds at time of application
11. Method of application
12. Pertinent comments regarding unusual test conditions or performance
13. Intervals between treatment and observation
14. Yields
15. Time of harvest
16. Percent control of specific pests compared to untreated controls and the commercial standard

These points are not new; they are well known; many good scientists have been following them. However, now it is essential to include them in the reports of your experiments if your scientific experiments are to be used supportively to obtain registrations. In this connection, it is very worthwhile to read the EPA guidelines and especially the appendixes where suggested experimental methods are described.

Today EPA judges efficacy from two points of view: effectiveness and usefulness. Effectiveness is defined as the ability of the product to control the pest. Usefulness is defined as the ability to apply the product according to directions on the label without causing unreasonable adverse effects on man, his food, or the environment.

THE REGISTRATION PROCESS

The Federal Insecticide, Fungicide, and Rodenticide Act is the law governing registration of pesticides. It is commonly called FIFRA. The present law was enacted by Congress in 1972 and amended slightly in 1975 and 1978. This is the basic charter for the EPA, as well as for those of us in industry who work to get products registered.

FIFRA is interpreted and enforced by EPA. In 1975 the EPA published regulations and proposed guidelines concerning data required to obtain product registration. Their official interpretations of the law are finalized through regulations or rules written by them and published in the *Federal Register.* These rules or regulations have the force of law.

Validation of previous scientific data appears to be a serious problem at the present time. This process is one in which EPA is carefully scrutinizing and evaluating all the data in their registration files to determine validity based on today's standards and technology. I would like to suggest that we apply the principle of our own Common Law which, going back to the British, says that a person is innocent until proven guilty — rather than the French Common Law that says the person is guilty until proven innocent. In other words, the scientific data in the EPA registration files should be accepted as valid. If there is good reason to question the validity of the data in some specific case or some new concept or knowledge arises that requires reexamination of old data, then it should be done.

Nevertheless, a reasonably liberal approach should be taken when examining the results found using older scientific techniques. As pointed out previously, often there is more than one way to design an experiment to ascertain the desired factual basis from which a scientific conclusion may be drawn. A tremendous amount of time, money, and manpower would be saved by assuming the data are acceptable unless there is good reason to believe otherwise. After all, it was acceptable when submitted, and it was critically examined at that time. In any case, this is a fundamental point, and we should prevent this situation from arising again in the future. Scientific techniques are in a constant state of flux, and 5 or 10 years from now, the same questions concerning validation of today's data could again become a problem.

New scientific discoveries frequently cast our problem in a new light or evoke the need for a new interpretation of the regulations. This is merely one more reason why the regulations must be general and the guidelines should be only suggestions on how data may be obtained. Scientists in industry, academia, EPA, and other parts of government must be permitted to use their scientific judgment and not be compelled to justify every decision by a check-box approach of prescribed data gathered according to a cookbook recipe.

At present Congress, industry, farmers, and environmentalists are dissatisfied with regulations that are being pyramided to plug doubts, oversights, and gaps in earlier regulations. The law is relatively simple: let's not make it so cumbersome with rules and regulations that it becomes too ponderous to work efficiently. The general problems have been defined by law. The solutions will not come through regulations per se, but by scientists in EPA, industry, and academia using the best available technology and working together.

It is well known that our analytical techniques are so good these days that a zero tolerance is scientifically impossible. However, it should be possible to have a negligible residue tolerance. A negligible residue was considered to be 1/2000 of the no-effect level based on a 90-day feeding study and should not exceed 0.2 ppm. This is a very practical concept that has been used in the past and it should be reinstated.

Pesticide Application

The current law has significant points of difference from previous versions. Three of these points are of particular interest:

1. The pesticide must be applied in accordance with label instructions.
2. It is an unlawful act not to use the pesticide in accordance with the label instructions. Some of the major items on labels are the use or pest for which the pesticide is applied and the method of application, the dosage employed, the volume applied, and the precautions and warnings.
3. Criminal penalties — "In general any registrant, commercial applicator, wholesaler, dealer, retailer, or their distributor who knowingly violates any provision of this act shall be guilty of a misdemeanor and shall on conviction be fined not more than $25,000 or imprisoned for not more than 1 year in jail or both." There are civil penalties which carry less severe punishments.

FIFRA is interpreted and enforced by EPA. Official interpretations of the law are finalized through regulations or rules written by them and published in the *Federal Register.* These rules or regulations have the force of law.

The enforcement of this law is carried out by the Enforcement Division of EPA. The overall enforcement policies are determined in Washington, but the actual enforcement activities are carried out through the ten regional offices. It is noteworthy that these regions have a great deal of autonomy and discriminatory judgment as to their enforcement activities.

Pesticide Labels

It is important to know that every label for a product does not carry a recommendation for aerial application. Further, even on labels with recommendations for aerial application, there are major subject differences. For example, Esteron® 6E herbicide (Dow herbicide), a 6-lb/gal concentrate of a low-volatile, 2,4-dichlorophenoxy (2,4-D) ester (isooctyl ester), carries a 1- to 5-gal volume recommendation for aerial application; i.e., the applicator may select a spray volume anywhere between 1 and 5 gal/acre in which to apply the desired amount of 2,4-D.

Tordon® 22K (Dow herbicide) weed killer has a 1- to 4-gal aerial spray recommendation, but Esteron® 99 (Dow herbicide) concentrate, a 4-lb/gal low-volatile 2,4-D ester formulation, carries a 3- to 5-gal spray volume recommendation for aerial applications. Thus, with Esteron® 99 concentrate, the pound dosage of 2,4-D used must be diluted to 3 to 5 gal of finished spray before application by aircraft.

Formula 40® and DMA® 4 (Dow herbicides) herbicides are both 4 lb/gal 2,4-D amine formulations, but use different amines. The labels for these products carry recommendations for aerial applications, and for these products there is no volume limitation. The labels state in general terms, "Use adequate volume for uniform coverage"; no specific volume is specified. We believe this is an excellent recommendation. It provides the local professional like yourself with the latitude to determine the spray volume he can safely apply with his equipment and to fit his local conditions.

So far we have been discussing only general recommendations for aerial application and spray volume rates. In addition to these variations, there are also different use recommendations for Dow products. For instance, only our Formula 40® and Esteron® 99 concentrate herbicide products carry a recommendation for the late application of 2,4-D in corn to control weed seed production and to act as a harvest aid. Other 2,4-D labels do not carry this recommendation.

Common Use Recommendations

Why don't we have labels giving the common use recommendations? The answer is that we try to have such labels, but there are two difficulties. One is that the label has to be a general label applicable to the entire U.S. Even as an innovation appears most promising or seems obviously to be an impoundment, any large company still has to see if it will survive the test of time and Murphy's law. As you know Murphy's law states that "if anything can go wrong it will". There is no test like large-scale use over a period of time to find out improvements for your product at label directions.

The other difficulty is at least equally formidable. EPA has to approve the new use, the new practice, or the label change. This nearly always is a major problem as the agency requires detailed documentation proving that the change works and that it does not cause an adverse effect on the environment. The only exception to this is if you wish to make the label more restrictive or add an additional precaution or warning. This type of label change may be easily done.

The difficulty in making a label change may be illustrated with the label for Esteron® 99 concentrate. Last year after some effort, we finally obtained a new use recommendation which was to control wild garlic after harvest in stubble grain fields. Just after this was issued, the Illinois Extension Service pointed out that this same wild garlic problem exists in soybean fields. They requested the word soybeans be added after the small grains and corn specified in the original text. Our amended application pointed out that this was essentially a very minor change, practically of an editorial nature that should have been in the original application. We also made reference to the original application for efficacy data, etc. Nonetheless, our request was turned down by the agency with the comment "You have not provided adequate data to support this proposed amendment. Please refer to the regulations implementing Section 3 of the act; specifically 162.8 and 162.11 (40 FR 28272 July 3, 1975)."

Section 162.8 covers data requirements for new registrations, i.e., efficacy including effective dosage ranges, phytotoxic effects on crop, general chemistry, environmental chemistry, product hazard, hazard to humans and domestic animals, hazard to non-target organism, acute toxicity data, subacute and chronic toxicity data, etc. Section 162.11 deals with hazards to wildlife, etc. In other words, all the data is being requested as if this were the first use registered for a new, unknown compound. This may sound absurd, but it is the climate in which we must operate today. It also points out the difficulties in getting a new use registered.

One other point concerning labels is that when you have many 2,4-D products registered, it is impractical to change them all at once. Therefore, there are always differences in our product labels, although we try to keep them at a minimum. These labels are developed by evolution over a period of time by experience and market need. This is why there are differences in our phenoxy labels for aerial application.

One suggested interim protocol for data to support label recommendations for aerial application of less than 3 gal/acre has been developed by EPA. This is very detailed protocol. The concluding statements read as follows.

"Where the extent of a potential drift hazard can be adequately defined by relating the bioassay results to the physical data generated in the drift tests, label recommendations, directions for use and limitations will correspond to the appropriate application conditions, techniques, equipment and adjuvants used in drift testing and will include a minimum safe distance (buffer zone) between desirable susceptible nontarget plants and the area to be treated. The various application conditions, techniques, equipment and adjuvants may require a corresponding variety of buffer zone dimensions to fit the situation. To support a specific safe distance off-target, bioassy plants must demonstrate the no-effect level at one half this recommended distance. Label directions, recommendations and limitations must prescribe the limiting factors under which this margin of safety may be achieved."

The final sentence in this suggested protocol is "Where a potential drift hazard exists, but is not evaluated, registration will be denied."

Another problem is that many of the present aerial application practices were developed empirically over a long period of time. Thus, data in the form of scientific articles or formal reports on these practices often do not exist. Further, until the current law was passed, detailed documentation proving the safeness as well as the efficacy of the change was unnecessary.

Area-Specific Situations

Another difficulty that has arisen is that information that we have developed through our research in one area of the country sometimes has not been accepted by EPA as being applicable in another region. For example, EPA is concerned that drift data for aerial applications of Texas rangelands are not suitable to support safe aerial uses in say the small grain areas. We have tried unsuccessfully to say that this could be handled by careful phraseology on the label.

Dow has been working hard trying to obtain low-volume application rates, particularly for small grains. Recent attempts took place September 23, 1976 when a special meeting was held with EPA personnel on the subject of low-volume aerial application of phenoxy herbicides in the grain belt area of North and South Dakota, Montana, and Minnesota. Representatives for university and extension service personnel, industry, and aerial applicators were present.

We are positive the message got across that for many years low-volume application of herbicides has been a safe and common practice in the grain belt. However, the jury is still out as to whether EPA will permit us to revise a phenoxy label and put on it an aerial recommendation of less than 3 gal spray volume per acre.

In the past, Dow has been actively engaged in aerial application research that has resulted in the present label recommendations. It is apparent that even maintaining as well as obtaining such labels for aerial application is a significant expense that has to be reflected in the prices of Dow products. This is merely one area where there are increased costs that have to be passed on to the end user. The increases are too great for the chemical industry to absorb in today's economy.

From the foregoing, it is evident that industry strongly supports aerial application of agricultural chemicals. However, careless use of this technique results in adverse publicity. There have been a sufficient number of careless aerial applications contrary to label directions that can do nothing but provoke further legislative action.

Damage to the environment is king today. The publicity that could result from one poor aerial application of 2,4-D or any other agricultural chemical could seriously affect all of us. It is noteworthy to compare the publicity resulting from different types of events that could become news. Consider the news coverage of a recent oil spill off the coast of New Hampshire vs. the deaths of at least 650,000 people that were killed in a recent earthquake in China. Everyone knows all about the oil spills, but how many know anything at all about the deaths of 650,000 people? It is imperative that all label directions be followed and the utmost care taken to apply herbicides safely. If not, the priviledge of aerial application of agricultural chemicals could be lost or greatly restricted.

Labels for aerial application of 2,4-D herbicides vary for different products. This must be kept in mind because the law today requires strict following of the directions on the label and provides severe penalties for violations.

Aerial application can be a very safe and efficient method of applying herbicides, but it has to be done in a responsible manner by responsible people. Misapplication of herbicides by aerial applicators will cause legislation to make this method illegal or at least severely restrict its use. The most likely misapplications are caused by bad pilot judgment such as applying under unfavorable conditions or using faulty equipment.

REBUTTABLE PRESUMPTION AGAINST REREGISTRATION

The EPA was formally scheduled to review the widely used weed and brush killer 2,4,5-trichlorophenoxy (2,4,5-T). We should examine this review process — the Rebuttable Presumption Against Reregistration (RPAR) — and its significance to U.S. agriculture.[6,7]

Notice of an RPAR means that EPA believes it has found some evidence of a potential health hazard and is asking producers and users of the product to comment on this evidence and describe counter-balancing economic benefits. After the submission of this information, the Agency will decide whether the compound is reasonably safe as used, whether additional limitations are needed, or whether it should be removed from the market.

RPAR is only one of several actions the EPA Administrator can take to control agricultural chemicals. The most severe is an Emergency Suspension — when the Administrator determines there is an imminent hazard from continued use of a chemical. Decreasing in severity, the other steps are suspension with hearing, notice of intent to cancel registration, notice of intent to reclassify, notice of intent to hold a hearing; and the least severe — RPAR.

It is significant that the 2,4,5-T RPAR is an inquiry proceeding which is merely calling for the submission of evidence supporting and rebutting claims of risk and benefit. It does not have the urgency of a suspension or cancellation. In fact, EPA has estimated that approximately 60% of the herbicides and pesticides presently registered for use in the U.S. will be reviewed under the RPAR process.

2,4,5-T has compiled an enviable safety record in the U.S. since 1948, controlling unwanted brush on rangeland, controlling trees and brush on highway and utility rights-of-way, limiting weeds in rice paddies, and in forest management. The facts which will be gathered in this RPAR review will determine the extent of its future use.

Special interest groups which oppose certain timber management practices, or the use of virtually any chemical in the environment, have been exerting strong pressure to have 2,4,5-T banned. The RPAR review provides an excellent opportunity to resolve the issues they have raised. These groups fail to understand that so far there is no alternative-chemical or nonchemical that achieves the public benefits of 2,4,5-T.

2,4,5-T has been used for 30 years to control trees, brush, and weeds which (1) threaten the growth of rice, conifer seedlings, or rangeland grass and (2) create hazardous situations on road shoulders and utility rights-of-way. During this period, the toxicity of 2,4,5-T has been exhaustively studied by scientists world-wide. The consensus of these scientists is that it is not hazardous to human or animal life under normal conditions of use and even under conditions of substantial misuse.

Recent debate has centered on the presence of trace amounts (less than 1 part in 10 million) of a very toxic contaminant, tetrachlorodibenzo-*p*-dioxin (TCDD), in 2,4,5-T. However, toxic TCDD may be, at the levels present in commercial preparations, it does not contribute to the toxicity of 2,4,5-T. The amount of TCDD distributed in 2,4,5-T in the U.S. is probably not more than 8 oz annually, spread over 5 million acres. Further, its rapid rate of degradation in sunlights helps to explain why it has not been found to be accumulating in the environment. The registered uses for 2,4,5-T are not exposing the public, or the environment in general, to toxic levels of TCDD or of 2,4,5-T itself.

2,4,5-T Dispute Resolution Conference

We recognize that modern U.S. society is faced with the need to make decisions about a burgeoning number of complex public interest problems, such as whether, how, and under what circumstances to use chemicals, nuclear power, computers, or other technology. The answers to these questions are neither simple nor absolute. They depend on "quality of life" decisions about how one wishes to live on this planet.

One common feature of all these problems is the need to make decisions about esoteric scientific matters. Traditionally, science seeks to avoid such decisions until all the necessary data, research results, and opinions have been assembled, evaluated, and judged so that a scientific "consensus" can be reached.

We no longer have the luxury of awaiting a final scientific consensus in this traditional sense. Decisions must be made now. There is no other alternative. We either do or do not use a chemical; we either do or do not use nuclear power. To await the final, traditional scientific consensus means that the barn door may have been closed long after the animals have escaped.

In a free and democratic society, ultimately the public must make the final decisions on quality of life issues. Scientists may debate chemical hazards; legislators may evaluate them; administrative agencies may examine them; courts may adjudicate them. However, ultimately the public must decide the critical issues.

Responsive public decision-making in a free and democratic society necessarily requires that the public be furnished the best available information in a form that can be understood and used. Science and scientists thus have an obligation to furnish the public their best possible judgments with regard to scientific issues. Scientists are aware that the public perception of risk varies with the nature of the hazard. Science cannot make judgments for society, but science can provide the information that permits more informed judgments.

To this point, science has not been able to furnish the traditional scientific consensus for decision-makers because the necessary data for such a traditional consensus are not yet available. This means that the advocates of opposing interests each present their data and their conclusions in an effort to "win" for the result they seek. The public recognizes such contentions as adversarial ones. It is unable to evaluate the complex evidence furnished, but has nowhere to turn for independent scientific judgment. To this point, the result has been loss of credibility, disaffection, and growing concern about whether our societal institutions are adequate.

The independent scientists who were assembled from throughout the free world and the observers who watched and commented are convinced that decision-makers in our society and finally the public will be furnished the best available credible scientific evaluations that it is possible to furnish, even where the traditional scientific consensus is not yet possible because so much remains uncertain and unclear.

Scientists have come to participate as scientists, dedicated to science and the service of mankind. They have been able to arrive at consensus with regard to most of the important and controversial issues that have been addressed in this conference. They believe that their pilot effort in developing an effective dispute resolution mechanism represents a contribution to the process by which our society may come to judgments with regard to scientific matters.

The conveners of this conference have long believed that our methods for resolving socioscientific disputes are inadequate and that new methods of dispute resolution must be developed and tested. Academic intellectual pursuit is not enough. A specific, controversial subject to dispute, which has not responded satisfactorily to existing mechanisms, was chosen as the conference topic. It provides almost the perfect model. It has been in bitter controversy for well over a decade. It has been the subject of recurrent scientific advisory committees, legislative, regulatory, executive, administrative, and judicial action on state, federal, and local levels — more than any other single discrete product. Little seems to have escaped repeated trial as a dispute resolution mechanism, to this point with little success in achieving satisfactory resolution.

2,4,5-T first came to active public consciousness as a result of its use as a defoliant in Vietnam. The early concern with regard to its potential harmful effects was colored by the acrimony and bitterness of our Vietnam involvement. In 1969 the discovery of TCDD, a highly toxic contaminant in 2,4,5-T came to the fore. As Vietnam faded into the background, other concerns about carcinogenicity and teratogenicity maintained high public interest. Books, feature news articles, national television specials, and other continuing publicity recently focused public concern on the possibility that 2,4,5-T caused abortions (Alsea, Ore.) and a variety of ills, including cancer, to veterans of the Vietnam conflict.

The scientific dispute resolution mechanism proposed and tested in Washington, D.C., June 4, 5, and 6, 1979, was one which brought together qualified scientists throughout the world who had information or expertise with regard to 2,4,5-T issues. The purpose was to see whether in the presence of each other, after being furnished the plethora of available information, after freely exchanging data and their interpretations of it, and after examining conflicting views, these scientists could arrive at a consensus with respect to the available data. To ensure credibility, this was to be done in the presence of "observers" from parties at interest, the media, and from the public.

The Research Foundation of the American Farm Bureau Federation, using member and industry contributed funds, provided the financial support for the conference. One of the critical key conditions of accepting financial support at the inception and throughout has been that there be no effort to influence the conference procedures employed or the selection of participants. There has been no such influence.

Using this model, invitations were extended to known interested persons or organizations in government, academia, industry, environment and consumer groups, and to foreign scientists who had been involved in aspects of the relevant scientific inquiries.

It is testimony to the dedication of scientists to the public interest and their conviction that something new is needed, that the response was immediate and constructive. A total of 48 scientists attended from 20 states of this nation, as well as 11 from the nations of New Zealand, Switzerland, Italy, Germany, France, Sweden, and Canada. Many came because an effort was underway to test a new mechanism for resolving important socioscientific problems. In addition to these scientists, 50 observers attended from government, academia, environmental organizations, industry, the media, and even the U.S.S.R.

SUMMARY AND CONCLUSIONS

As scientists, we would far prefer to be able to wait until tomorrow before we express our scientific views. As members of society, however, we know that we cannot wait until then. We have an obligation to let the public know today of our conclusions based on evidence available, so that they may come to decisions for themselves. We are aware of our scientific limitations. We are primarily experts in biology, chemistry, statistics, and economics. Other disciplines must be involved in the fashioning of any long-range dispute resolution mechanism of more general application.

Not only would this include scientists from other "hard" disciplines, such as nuclear physics and computer technology, but from the "soft" sciences, including sociology, law, government, political science, and the like.

Science is neither good nor bad. It is how science is used by society that counts. We believe that science can help make this world a better place in which to live. We hope that society will learn how to employ science to its benefit. We as scientists will contribute our valiant effort to that end. We hope that others will join with us in the pursuit of the goal of a better life for all.

REFERENCES

1. CFR 40 Part 162, Regulations for the Registration, Reregistration and Classification of Pesticides of the Federal Insecticide, Fungicide and Rodenticide Act, U.S. Environmental Protection Agency, Washington, D.C. 1974.
2. General Instructions for Registering Pesticide Products under the Amended FIFRA, U.S. Environmental Protection Agency, Washington, D.C., 1976.
3. Registration of Pesticide in the U.S., U.S. Department of Agriculture, Health, Education, and Welfare, Food and Drug Administration, U.S. Government Printing Office, Washington, D.C., 1968.
4. EPA Notice to Producers, Formulators and Registrants, PR Notice 78-5, U.S. Environmental Protection Agency, Washington, D.C., 1978.
5. EPA, The Federal Insecticide, Fungicide and Rodenticide Act, As Amended, U.S. Environmental Protection Agency, Washington, D.C., 1978.
6. EPA, Rebuttable presumption against registration and continued registration of pesticide products containing 2,4,5-T, *Fed. Regist.*, 43 (78), 17116, 1978.
7. EPA, Decision and emergency suspension order suspending registrations for the forest; rights-of-way, and pasture uses of 2,4,5-T, *Fed. Regist.*, 44(52), 15874, 1979.
8. **Bowey, R. W. and Young, A. L.,** *The Science of 2,4,5-T and Associated Phenoxy Herbicides,* John Wiley & Sons, New York, 1980.

Chapter 5

REMOTE SENSING TECHNOLOGY FOR RIGHTS-OF-WAY MANAGEMENT*

INTRODUCTION

The Earth Resources Observation Systems (EROS) Program of the U.S. Department of the Interior (USDI), administered by the U.S. Geological Survey (USGS) was created in 1966 to apply remote-sensing techniques to the inventory, monitoring, and management of resources. To meet its primary objective, the EROS Program includes research and training in the interpretation and application of remotely sensed data and provides remotely sensed data at nominal cost to scientists, resource planners, managers, and the public.[1,2,4]

The EROS Data Center, outside Sioux Falls, S.D., is operated by the EROS Program to provide access to the National Aeronautics and Space Administration's (NASA) LANDSAT [formerly Earth Resources Technology Satellite (ERTS)] imagery, aerial photography acquired by the USDI and photography and imagery acquired by NASA from research aircraft and from Skylab, Apollo, and Gemini spacecraft. The primary functions of the Data Center are data storage and reproduction, and user assistance and training.

At the heart of the Data Center is a central computer complex which controls a data base of several million images and photographs of the Earth's surface features, performs searches of specific geographic areas of interest, and serves as a management tool for the entire data reproduction process. The computerized data storage and retrieval system is based on a geographic system of latitude and longitude, supplemented by information about image quality, cloud cover, and type of data. A customer's inquiry about availability of remotely sensed data may be about geographic point location or a rectangular area specified by latitude and longitude corner coordinates. Depending on customer requirements, a computer geographic search will print out a listing of available imagery and photography from which the customer can make final selection.

Guidance in the use of remotely sensed data is available at the EROS Data Center in the form of scheduled training courses and workshops. The scientific teaching staff at the Data Center periodically offers discipline-oriented courses in subjects such as agriculture, forestry, geography, geology, and hydrology. Visitors to the Data Center will also get assistance in the operation of specialized equipment such as densitometers, additive color viewers, zoom transfer scopes, and stereo viewers, and in the use of computerized multispectral systems to classify specific phenomena.

The EROS Data Center also operates several Applications Assistance Facilities which maintain microfilm copies of data archived at the Center and provide computer terminal inquiry and order capability to the central computer complex at the EROS Data Center. Scientific personnel are available for assistance in applying the data to a variety of resource and environmental problems and for assistance in ordering data from the Data Center.

EROS Data Reference Files have been established throughout the U.S. to maintain microfilm copies of the data available from the Data Center and to provide guides to assist the visitor in reviewing and ordering data. This allows the visitor to view microfilm copies of the data before placing an order. Applications assistance by scientists is not provided at EROS Data Reference Files locations.

* By C. M. Schwartz (deceased), Office of the Chief of Engineers, Washington, D.C.

LANDSAT (EARTH RESOURCES TECHNOLOGY SATELLITE) DATA

The first Earth Resources Technology Satellite, ERTS-1 (now renamed LANDSAT-1), was launched July 23, 1972. LANDSAT-2 was launched on January 22, 1975. LANDSAT flies in a circular orbit 570 mi (920 km) above the Earth surface and circles the Earth every 103 min, or roughly 14 times a day. Each daytime orbital pass is from north to south. From such a vantage point, each LANDSAT can cover the entire globe, except for the poles, with repetitive coverage every 18 days. A unique feature of the satellite, because of its orbit, is that it views the Earth at the same local time, roughly 9:30 a.m. at the equator, on each pass. The sensors on board the spacecraft transmit images to NASA receiving stations in Alaska, California, and Maryland, either directly or from data stored on tape recorders. The data are converted from electronic signals to photographic images and computer compatible tapes at NASA's Goddard Space Flight Center (GSFC) in Greenbelt, Md. Master reproducible copies are flown to the EROS Data Center in Sioux Falls, where images are placed in the public domain and where request for reproductions are filled for the scientific community, industry, and public at large. Because of the experimental nature of the satellite and the limited capabilities of NASA ground processing equipment at Greenbelt, approximately 30 days are required from the time the signals are first received on the ground to the time the data is available to the public at the EROS Data Center.

LANDSAT presently carries three data acquisition systems:

1. A multispectral scanner (MSS) (four spectral bands)
2. A return beam vidicon (RBV) or television system
3. A data collection system (DCS) to relay environmental data from ground-based data collection platforms (DCPs)

The MSS is the primary sensor system and acquires images of 115 mi (185 km) per side in four spectral bands in the visible and near-IR portions of the electromagnetic spectrum. These four bands are

1. Band 4, the green band, 0.5 to 0.6 μm, emphasizes movement of sediment-laden water and delineates areas of shallow water, such as shoals, reefs, etc.
2. Band 5, the red band 0.6 to 0.7 μm, emphasizes cultural features
3. Band 6, the near-IR band, 0.7 to 0.8 μm, emphasizes vegetation, the boundary between land and water, and landforms
4. Band 7, the second near-IR band, 0.8 to 1.1 μm, provides the best penetration of atmospheric haze and also emphasizes vegetation, the boundary between land and water, and landforms

MORPHOLOGICAL ANALYSES

An analysis of the four individual black and white images or the false-color IR composite images often permits scientists to identify and inventory different environmental phenomena, such as distribution and general type of vegetation, regional geologic structure, and a real extent of surface water. The repetitive (9 or 18 days) and seasonal coverage provided by LANDSAT imagery is an important new tool for the interpretation of dynamic phenomena. It should be noted that because of the Earth rotation and the fact that the image is created by an optical-mechanical scanner, ERTS MSS images are parallelograms, not squares. The sides are parallel to the orbital track of the satellite on the Earth surface. The arbitrary forward overlap between consecutive

LANDSAT images is approximately 10%. The sidelap between adjacent orbits ranges from 14% at the equator to 85% at the 80th parallel of latitude.

Latitude and longitude tick marks are depicted at 30-min intervals outside the image edge. These geographic reference marks are noted in degrees, minutes, and compass direction. A 15-step gray scale table is exposed on every frame of LANDSAT imagery as it is produced. This scale is used to monitor and control printing and processing functions and to provide a reference for analysis related to a particular image. The annotation block directly over the gray scale contains data that give the unique image identification, geographic location, and time (with respect to Greenwich mean time) the image was acquired.

Band 5 provides the greatest information for a single black and white image. This band usually gives the best general-purpose view of the Earth surface. By ordering a complete set of black and white images from all four bands, however, it is possible to examine how a particular area differs in appearance when filtered to green, red, and near-IR wavelengths. In MSS false-color images, healthy vegetation appears bright red rather than green; clear water appears black; sediment-laden water is powdery blue in color; and urban centers often appear blue or blue-gray. MSS false-color composite images which have not already been prepared can be ordered from the Data Center, but carry a one-time initial preparation charge of $50, not including the cost of any products ordered for the resulting composite.

A set of LANDSAT images has been prepared for the conterminous U.S. The 470 scenes required to cover the U.S. are available in a single black and white band (band 5), all four bands of black and white, or high-quality color composites. The scenes selected were chosen on the basis of quality, optimum time of year (generally spring or summer), and minimum cloud cover.

LANDSAT data in digital form is available as Computer Compatible Tapes (CCT). These tapes are standard 1/2-in.-wide (12.7-mm) magnetic tapes and may be requested in either seven- or nine-track format at 800 or 1600 bpi. Four CCTs are required for the digital data corresponding to one LANDSAT image. The data for the four bands are interleaved among the four tapes thereby necessitating all tapes to complete a set.

GEOGRAPHIC SEARCH AND INQUIRE SYSTEM

Request for information about imagery of a specific area will initiate a computerized geographic search. The search can be initated by mail, visit, or phone to either the EROS Data Center or one of the EROS Applications Assistance Facilities. You may request a geographic search using any of the three following options:

1. Point search — all images or photographs with any portion falling within 50 mi (80 km) of the point will be included.
2. Area rectangle — any area of interest defined by four corner coordinates of latitudes and longitudes. All images or photographs with any coverage of the area will be listed. The area must not exceed 200 1° squares (for example, 10° latitude by 20° longitude).
3. You may enclose a map with a point or area indicated.

When requesting a geographic search from the Data Center, all relevant information must be provided. This should include acceptable dates and seasons, type of imagery preferred, color, false-color IR, or black and white IR, cloud cover, and quality. Cloud cover is given only in percentage, hence, no assurance can be given as to where clouds will appear on the resulting photographs or images. A description of your intended application and the use of the data will assist the researcher at the Data Center who initiates the search, thereby resulting in a more concise response to your inquiry.

Geographic areas must be clearly identified and should be limited in size as much as possible to avoid a potentially long computer listing and the need to review large numbers of choices. Latitude and longitude coordinate specification is preferred, since this is the method required for the computer geographic search.

Specification in degrees and minutes normally provides sufficient location accuracy. (Each degree of latitude or longitude is divided into 60 min, and each minute is divided into 60 sec; 1 min of latitude is roughly 1 mi.) Coordinates can be found for your area of interest from maps in a home library, atlas, or on many state road maps. The alternate method is to enclose a map with your area specifically outlined.

THE COMPUTER LISTING SYSTEM

The computer listing will contain all images and photographs available over or close to your area of interest. Two printed lines on the computer listing describe in detail the characteristics of each image or photograph. A number of entries may be listed, depending upon the size of the area selected and the restrictions of the supplemental data. Imagery or photography may be available from more than one source — LANDSAT, Skylab, NASA aircraft, or survey aircraft — all of which differ significantly in sensor or camera characteristics. Thus, each entry on the computer listing should be carefully studied to determine the best selection for your interest and application. The first line of each entry on the computer listing gives data characteristics, along with information required for subsequent ordering. Specific items to be considered are

1. Type cover — indicates type of coverage from one of the four sources of photographs or images available from the Data Center (this must be used to determine the correct order form and ordering procedure)
2. Scene identification — indicates a 15-character unique order number to be transferred to the order form when placing an order
3. Film source — indicates black and white, color, false-color IR, or black and white IR film
4. Quality — indicates the quality of master film roll on the following scale: 0 to 2, very poor; 5 to 6, fair; and 7 to 9 good to excellent
5. Cloud cover — indicates the approximate cloud cover in percentage, that is 10, 40, 80%, and so on
6. Date acquired — gives the date the exposure was made, read as year, month, day, e.g., 740318 indicates March 18, 1974
7. Center point — gives the latitude and longitude coordinates of the center point of the scene in depgrees, minutes, and seconds
8. Band availability — "1" indicates availability of a specific spectral band; " 0" indicates nonavailability

For all aircraft photographic accessions these items are also noted:

1. Type accession — indicates whether the computer is describing a scene (single photograph), strip (a series of photographs in a strip), or a photoindex (mosaic of overlapping photographs) (the pertinent codes are "5" for an individual photograph, "2" for a photostrip accession, and "1" for a photoindex)
2. First frame-last frame — indicates, for all strip accessions, the first and last frames on the master file roll (if the same number is listed, a single photograph is described)

3. Number of frames — lists, for your convenience, the number of frames in the film between the first and last frames

The remaining information on the first line of the computer listing is included primarily for technical use or internal use by the Data Center. The second line of each entry contains the corner coordinates in latitude and longitude which may be plotted to determine the exact ground coverage of either individual frames, strips, or photoindexes.

The image size of the master reproducible is shown (in millimeters) at the end of the second line and may be needed to order some types of aircraft data. A template is provided for decoding the computer listing. When a choice of the most suitable data has been made, place an order using the appropriate order form.

Placing an Order

Orders for reproductions of data from the EROS Data Center are accepted from individuals, governmental organizations, universities, and industries in the U.S. and foreign countries. Orders may be placed by personal visit, telephone, or mail to the Data Center. Orders may also be placed at any of the EROS Applications Assistance Facilities.

Quality of the reproduction cannot exceed that of the master film available at the Data Center. Every product leaving the Data Center undergoes a rigid inspection to ensure that internal quality standards are met. Occasionally, you may find that a product has certain defects, such as small scratches, pinholes, or stains, or that the color balance or density of the reproduction is not exactly as you expected it should be. In these cases, you may be assured that it has been inspected against the quality of the original reproducible and that everything has been done to produce as good a product as possible. As with any large production facility, however, substandard products sometime reach the user. Such products should be returned to the Data Center for reprocessing or a refund.

Applications Assistance

Training in remote sensing and assistance in techniques for analysis for remotely sensed data are available at the EROS Data Center. Inquiries can be made by telephone, letter, or personal visit. The Applications Assistance Branch of the Data Center should be contacted before making travel plans.

Periodically, training programs in remote sensing are given at the EROS Data Center. Normally these programs are up to 1 week in length and stress the use of data in a particular application, such as agriculture inventory of water management. Approximately twice a year, a 3-week course is offered for foreign nationals that stresses the fundamentals of remote sensing and introduces the application of remotely sensed data to different resource problems. The formal training is supplemented by a series of slide and magnetic tape packages that cover the basic methodology of remote sensing and selected applications.

The EROS Data Center has a developing capability to perform computer-assisted analysis of imagery. Special devices permit the experimental use of digital analysis techniques to classify phenomena by their reflectance or emittance in different parts of the electromagnetic spectrum.

DIGITAL IMAGE ENHANCEMENT SYSTEM

The USGS's EROS Data Center produces on LANDSAT image product referred to as computer enhanced imagery. This product reflects advanced digital image processing techniques that perform certain geometric and radiometric corrections and perform contrast enhancement and edge enhancement. The enhanced image product is produced from digitally preprocessed LANDSAT CCTs on a laser beam film recording system. The processing philosophy and techniques were developed at the USGS Center of Astrogeology in Flagstaff, Ariz., and they have been modified by the EROS Data Center in Sioux Falls, as the Digital Image Enhancement System (EDIES).

The EDIES performs the following seven specific operations:

1. Radiometric restoration (destriping)
2. Sampling geometric restoration
3. Earth rotation correction
4. Detector misregistration correction
5. Contrast enhancement
6. Edge enhancement
7. Synthetic line generation

Detailed descriptions of these operations and comments are presented in the following paragraphs.

Radiometric Restoration

For each spectral band, the LANDSAT MSS scans six successive lines at a time using six individual detectors. Due to the differences between the radiometric transfer functions of these detectors, the resulting images often exhibit a striping effect which may interfere with the spectral classification accuracy and/or the pictoral quality of the image. To remove the striping effect, a histogram equalization technique is used.

Letting (M^i,S^i, i – 1 to 6) represent the image mean and SD corresponding to each detector, where M and S represent the desired mean and SD for the entire image, the destriping process can be defined as:

$$X_i\ (L, P) - X_i\ (L, P) \times \frac{S}{S} + \left(M - M_i \frac{S}{S_i}\right)$$

Where X_i (L,P) is the pixel value of the point (L = line, P = pixel) scanned by detector i. X_i (L,P) is the new pixel value after destriping. Globally, this model performs well. However, in some subregions where the local statistics are quite different from the global statistics, the destriping algorithm is not very effective. Thus, a local operator is applied as a second step to eliminate the residues.

The algorithm that is implemented to remove these residues provides filtering or smoothing. Each image is processed in groups of six lines, the first line of which is chosen as a reference and left unchanged. Each of the remaining five lines is processed to be similar to the line above. About each point in the line being processed, a local average is computed and compared to the corresponding local average in the line above. The magnitude of the difference between the local average in the line above and the local average in the line being processed is then compared to the modification limit which is provided as a parameter. The minimum of these two quantities, signed with the same line as the difference, is added to the point being processed. In addition, before being allowed to contribute to a local average, each point is subjected to a test. If the point differs from the corresponding point in the processed line above by more than an edge threshold, also supplied as a parameter, the point is replaced by the average of it and the point above.

Sampling Geometric Restoration

The scanning of a line of LANDSAT imagery is accomplished by an oscillating mirror that intercepts the radiation path. The angular velocity of the mirror is not constant, and the sampling rate is very uniform. As a result, the samples in the line are unevenly spaced and are displaced from the normal regularly spaced locations.

The correction is achieved by piecewise linear interpolations. Each scan line is divided into eight segments ($S_1, S_2, \ldots S_8$), and each segment, $S_i, 1 \leqslant i \leqslant 8$, is partitioned into n_i pixels with center coordinates $X_1, X_2, \ldots X_{n_i}$ and pixel values $X_1, X_2, \ldots X_{n_i}$, respectively.

Letting the ground error of segment S_i equal d_i pixels in length, the segment S_i is repartitioned into $n_i + d_i$ equal parts with center coordinates $Y_1, Y_2, \ldots Y_{n_i}$. For each Y_i, $1 \leqslant i \leqslant n_i + d_i$, a new pixel is assigned and its pixel value Y_i can be expressed as:

$$Y_i = \wedge X_j + (1 - \wedge) X_j + 1 \qquad 0 \leqslant \wedge \leqslant 1$$

Where X_j and $X_j + 1$ are two center coordinates of pixels which are adjacent to Y_i and $\wedge$ is the distance between Y_i and $X_j + 1$.

The process is repeated for all eight segments. All the new pixels form a mirror speed corrected line. The ground displacement error values are provided by GSFC.

Earth Rotation Correction

Skew distortions are introduced as a result of the Earth rotation and satellite movement in orbit as an image is acquired. The amount of skew present in a scene is largely a function of latitude and spacecraft heading. Each image is deskewed with an algorithm which translates the scan lines to the right by a calculated number of pixels dependent upon the associated Earth latitude such that terrain features are in proper position relative to each other from top to bottom of the scan.

Detector Misregistration Correction

Due to the physical detector alignment and sampling procedure utilized in the LANDSAT spacecraft, geographic data is misregistered by two pixels band-to-band. This misregistration is corrected by the insertion of an appropriate number of "one's" pixels in each scan line of each band to effectively shift the scan line into band-to-band registration. This is required, primarily, for the color film products.

Contrast Enhancement

In this technique, a contrast stretch is applied to LANDSAT imagery to provide optimal contrast and maximum color variation.

A linear stretch can be generated by mapping each gray level encountered in the picture, pixel-to-pixel, to the range (0,255) by the linear equation:

$$X(L, P) = \frac{X(L, P) - \text{Min} \times 225}{\text{Max} - \text{Min}}$$

Where X(L,P) is the pixel value at point (L = line, P = pixel) and X′ (L,P) is the corresponding pixel value after the stretch. Min and Max in the equation represent the minimum/ maximum gray levels established for each image.

The minimum and maximum gray levels are derived from histogram information and verified by using Image 100 image processing software to minimize the loss of meaningful information.

Edge Enhancement

The technique for enhancing the high-frequency image components is to first low-pass filter the image by a spatial convolution (a weighted average). The high frequency components are then extracted and added back to the image.

$$\text{Edge enhanced image} = \text{image} + [\text{image} - f(\text{image})]$$

$$= 2 \times \text{image} - f(\text{image})$$

where f(image) is the local average in the neighborhood of the image and the size of the neighborhood is 11 pixels long.

Filtering to boost the high spatial frequency content, however, has the effect of producting a shadow of opposite polarity adjacent to any abrupt discontinuity in the imagery. In most cases, this artifact is difficult to observe and is not significantly disturbing. Even when the effect is apparent, however, the advantage of improved apparent sharpness generally compensates the disadvantage of the introduction of the artifact.

Synthetic Line Generation Algorithm

The data loss on LANDSAT imagery, due to transmission errors or tape errors, presents cosmetic defects on the film products. In general, it is desirable to remove these cosmetic defects (with the exception to the cartographer and the scientist to whom the geometric accuracy is extremely important).

The synthetic line algorithm calculates the SDs and mean values of the first six scan lines. By comparing the statistics, a good line is selected as the first reference line. Then, a 2 × 2 edge detector is applied to the reference line and the next scan line.

The detector moves horizontally across the scan lines one pixel at a time. Once a horizontal edge is registered, the scan line next to the reference line is replaced by the reference line. Otherwise it will be considered as the new reference line and the process is repeated.

Since a 2 × 2 edge detector is capable of detecting five different types of edges, the probability of the existence of a true horizontal edge of more than 150 pixels long, which can be detected by the 2 × 2 detector, is less than (1/5) 150. Therefore, the algorithm is very effective for any data segment loss of a length of more than 150 pixels and with an error probability less than (1/5) 150.

Film Product Generation

After the digital data has been processed with the above algorithms, the data for each LANDSAT band is printed on 241-mm (9.5 in.) black-and-white film with a laser beam film recorder (LBR). The LBR is configured to perform an aspect ratio correction and to linearly display the digital data over a 1.6 density range on the black-and-white prints at various scales, or color prints at various scales. Cibachrome color products are also available. All color products are prepared photographically from black-and-white film products generated on the LBR.

SKYLAB DATA

The 1973 through 1974 Skylab missions provided a limited amount of retrievable remote sensing data from an orbiting space platform. This data was a preview of future remote sensing that will possibly be available from the space shuttle program. The Skylab data used for this project was photographic and limited, whereas the LANDSAT imagery is multispectral scanner and repetitive. The main advantages of Skylab vs. LANDSAT was the increased resoltuion and diverse product types.

Skylab imagery is indexed and listed at the Data Center by individual scene. Each Skylab image, viewed from an altitude of approximately 270 mi, covers the following general areas depending on the camera system: the S190A system covers approximately 100 by 100 mi. The resolution of S190A photos is such that an object of about 150 by 150 ft can be identified. The S190B system covers approximately 68 by 68 mi. The resolution of S190B photos is such that an object of about 75 by 75 ft can be identified. The resolution of any scene depends on the contrast of the cultural and geological features as well as the atmospheric conditions.[3,4]

REMOTE SENSING APPLICATIONS GUIDE

Since the 1930s, remote sensing has played an increasingly important role with the Corps of Engineers and in other activities dealing with the natural environment. Initially, aerial photography was used as a tool in mapping, reconnaissance, and construction activities as well as specific war-related planning. New and more complex data needs have surfaced to meet the different missions of the Corps of Engineers. Regulatory surveillance, environmental impact assessment, evaluation of nonstructural alternatives to river basin planning, and general facility management are all important applications of remote sensing programs.[5]

The Remote Sensing Guide provides managerial and technical evaluation for most field activities within the Corps of Engineers and other natural resource organizations. The content of the guide and its appendixes are designed to span civil works and military programs by assisting the user in:

1. Identifying, planning, and conducting remote sensing applications
2. Locating relevant remote sensing imagery data, expertise, and services internal and external to the Corps of Engineers
3. Acquiring an understanding of fundamentals

As such, the Remote Sensing Guide is a ready made reference; a source of specific guidance on a broad topic basis. The contents have been purposely generalized to provide a framework for applying remote sensing to a broad range of data useful in the solution of specific problems.

SUMMARY AND CONCLUSIONS

A wide variety of remote-sensed imagery is available for study of the geographic areas of the U.S. and the world through the EROS Data Center. With such study, it is possible to delineate geographic and geomorphic features of a right-of-way terrain, and specific maps may be processed to bring out particular features that are important for management. This analysis can also be used to periodically review the vegetative management program as it relates to a specific right-of-way system.

REFERENCES

1. **Anon.,** The EROS Data Center USASINF-74-43, Department of the Interior, U.S. Government Printing Office, Washington, D.C., 1975.
2. **Anon.,** Data Users Handbook, General Electric Company, Beltsville, Md., 1975.
3. **Coiner, J. C. and Ungar, S. C.,** Ground truth management system to support Multispectral Scanner (MSS) digital analysis, *Proc. Am. Soc. Photogrommetry,* 43, 130, 1977.
4. **Short, I. M., Lowman, P. D., Jr., Freden, S. C., and Finch, W. A.,** Mission to Earth; Landsat Views of the World, National Aeronautics and Space Administration, Washington, D.C., 1976.
5. U.S. Army Corps of Engineers, Remote Sensing Symposium, Proceedings, Office of the Chief of Engineers, U.S. Government Printing Office, Washington, D.C., 1980.

Chapter 6

ENVIRONMENTAL IMPACTS OF PIPELINE RIGHTS-OF-WAY IN THE COASTAL ZONE*

INTRODUCTION

The purpose of this chapter is to examine the environmental impacts of pipeline construction and operation in the coastal zone. In general, there is a serious and real interest in assessing the environmental impacts of these activities as they approach the shoreline, more through the beach and sand dune areas, across the marshes and wetlands, etc. until they reach the storage or refinery areas. The primary interest is in the movement of crude petroleum through these pipelines as an integral part of a deepwater port facility, but many of the impacts also apply directly to the movement of refined petroleum products or natural gas. The possibility of accidental release of oil from the pipeline is also discussed.[1-8]

GENERAL CHARACTERISTICS OF THE COASTAL ZONE

The coastal zone is broadly defined as the area extending seaward to the limit of territorial space and landward to a distance which includes all land areas harboring activities directly dependent upon or influenced by the water.

In the U.S. this coastal zone contains about half of the population of the nation and most of the major cities. There are over 80,000 mi of shoreline in the U.S. coastal zone and many biologically important estuaries and wetlands.

Wetlands are those land and water areas subject to regular inundation by tidal, riverine, or lacustrine flowage. Generally included are inland and coastal shallows, marshes, mudflats, estuaries, swamps, and similar areas in coastal and inland waters. The wetlands produce a rich nutrient supply that supports a wealth of aquatic and semiaquatic life, both in numbers and varieties. This wealth of life — particularly the fish, wildlife, and higher plants — is of tremendous interest to man for its esthetic, sport, and food value. The higher animal and plant forms are in turn dependent upon the microscopic forms for their life support systems. The world of animal and plant life is a world of energy and material transformation and cyclic phenomena, as life and death are continued.

Wetlands have been used by numberless individuals for various purposes. When drained they have provided rich tracts of farm land; when filled they have provided sites for homes, industry, and commerce. They have been mined for building materials, oil, sulfur, salts, and metals, and on occasions, have served as depositories for ever-increasing amounts of liquid and solid wastes. They have provided millions of people with a playground to enjoy nature and find excitement in fishing or hunting and other recreation at the edge of the water.

During the early years when America was first settled, there was an estimated 127 million acres of wetlands. Because of their accessibility to water transportation, these wetlands were targets for development. The Federal Swamp Acts of 1849, 1850, and 1860 paved the way for the transfer of nearly 65 million acres of wetlands to 15 states from federal to state administration in order to expedite their drainage, mostly for agriculture purposes. In the last wetlands survey conducted by the Fish and Wildlife Service of the U.S. Department of the Interior (USDI), 1955, it was reported that only

* By C. G. Ash (retired), formerly with the U.S. Army Corps of Engineers, Washington, D.C.

an estimated 75 million acres remained. Of this total, 85% of the country's wetlands were inland freshwater areas, 2% were inland saline, 5% were coastal freshwater areas, and 7% were coastal saline areas. The smaller acreage of the coastal wetlands should not obscure the fact that they provide most of the food for marine life.

Marine Environment

Marine environments can be conveniently categorized into two major domains: the coastal ocean and the open ocean. The coastal ocean includes estuaries and adjacent wetlands, lagoons, the waters over the continental shelves, and many marginal seas. The open ocean is that part not significantly affected by its boundaries with the continents or by the shallow ocean bottom. There is no clear-cut boundary between the coastal and open ocean. Water, pollutants, and spilled oil can readily move from the coast to the open ocean or vice versa.

Open Ocean

Beyond the continental shelves, the ocean is relatively unaffected by its boundaries. There are no lateral restrictions, and the ocean bottom is some 4 to 6 km (2.5 to 4 mi) below the surface. Over most of the ocean, warm surface waters are separated from the cooler deep waters by the pycnocline, a rapid increase in density that more or less separates the surface oceanwaters from the deeper waters. While the deeper waters are known to move sluggishly after forming in the polar regions (primarily the Antarctic) and to return to the surface about 600 to 1000 years later, they do not affect the fate or movement of petroleum spilled on the ocean surface. Therefore, the movements of the deep ocean waters can be ignored, as it is the ocean surface circulation that is important.

Away from the land masses that interrupt surface water movements, ocean currents are primarily directed east-west. Only in the areas of the boundaries are the currents deflected to the north or south forming major boundary currents. Open ocean currents generally move the waters at speeds of a few kilometers per day. In the boundary currents such as the Gulf Stream or the Kuroshio Current, the waters move at speeds of 10 to 100 km/day.

Winds blowing across the ocean set the surface waters in motion. In the open ocean where tidal currents are relatively weak, these wind drift currents comprise about 40% of the surface currents and are probably the most important factor contributing to the movement of oil spills. An oil slick at sea will commonly move at a speed of about 3 to 4% of the wind speed. Because of these wind effects, oil spilled in the open ocean may be moved into the coastal ocean and eventually onto the shore or into wetland areas.

Coastal Ocean

The coastal ocean is the water adjacent to the shore; there the adjacent land boundary, freshwater runoff from the land, and local atmospheric conditions contribute significantly to the movement and mixing of the waters. The width of this zone is highly variable, and the outer boundary is not well defined. It may be quite narrow along coasts where the continental shelf is narrow and where oceanic conditions and the "permanent" currents come close to shore. Conversely, off a coast where the continental shelf is quite wide, the coastal ocean may be well over 100 mi wide. It is not, however, coincident with the continental shelf. Where the shelf is very narrow, the coastal ocean may extend beyond the shelf, and where the shelf is very wide, the coastal ocean may extend out from shore for only a fraction of the shelf width. The critical features defining this zone are the dominance of boundary effects and of transient local effects on the water movement.

There is considerable variation in the movement of the waters near the shore. Tidal currents are often strong; generally the strongest currents are near the shore. They generally parallel the coast and may be either oscillatory or rotary in character, depending on local geography. Inshore from the point where surface waves break, a longshore current that parallels the shoreline usually develops. The current direction depends upon the angle at which waves approach the shore. For example, along a coastline oriented in a north-south direction, with the ocean to the east, waves approaching the shoreline from the north-east will produce a longshore current that flows southward.

Outside the surf zone, local winds and the salinity gradient, caused by runoff from the land, combine to dominate the nontidal current in the near-shore zone. Since the waves breaking on a shoreline at any given time may have been for the most part generated by storm systems some distance away, the direction from which the waves approach the coast and the local wind direction are generally not correlated. Consequently, the direction of flow inside the surf zone may be opposite to the direction of the currents seaward of the surf zone.

Along coasts having relatively large freshwater inflows to the ocean, such as the Atlantic coast of the U.S., the salinity, and generally the density, for the water increases with distance offshore. The average net nontidal flow in the near-shore zone will, along such coasts in the Northern Hemisphere, be directed such that the shoreline is to the right of an observer looking downstream. Such currents are particularly well developed along coasts to the right (looking seaward) of the mouths of major estuaries — southward from the mouth of the Chesapeake Bay, for example.

At any particular time, however, the currents in the coastal ocean may be dominated by the local wind or by density-induced effects resulting from the wind. Hence, an offshore wind will transport surface waters offshore, particularly a wind blowing at an angle to the shoreline such that an observer with his back to the wind has the shoreline on his left. Under these conditions, warmer surface waters are transported offshore, and the cooler subsurface waters from offshore are brought to the surface near shore, a process known as upwelling. The resulting density distribution will produce, in the Northern Hemisphere, a current flowing along the coastline with the shore to the left of the current (looking downstream).

When such an offshore or longshore wind ceases, the warmer surface layers move shoreward. In this way, oil from an offshore spill can be carried onto the beach following the cessation of a wind that has produced an offshore movement of the oil slick. In the presence of an onshore wind, the wind-induced currents run parallel to the shore in a zone just outside the breakers.

Materials introduced into the near-shore zone will be transported away from the point of discharge by currents and will be mixed into the receiving waters by turbulence. Where there is continuous discharge of materials into the near-shore zone, a diverging plume (like the plume from a smoke stack) will extend downcurrent, with concentrations decreasing away from the source. Because of the changes in the current pattern, the plume will change direction, sometimes folding back on itself. More often, when the current direction approximately reverses, the remnants of the older plume will be carried along a path somewhat offset from the path of the new plume extending from the source in the new direction of the current. Over an extended period, a large volume of water with low oil concentrations will accumulate near the shore. Transient plumes of higher concentration plumes will also be superimposed over this water mass. The steady-state level of this general background concentration will depend upon the rate at which near-shore waters are exchanged with waters of the adjacent coastal or open ocean areas.

Estuaries

Most of the existing ports and harbors in the world are located on estuaries. Despite their variety of shapes and sizes, estuaries commonly exhibit similar physical and biological processes. To understand the fate and effects of petroleum when spilled in these areas, it is first necessary to consider these processes.

An estuary is a semienclosed coastal body of water that has a free connection with the open sea and within which seawater is measurably diluted with freshwater derived from land drainage. Four types of estuaries are recognized: drowned river valleys, fjord-like estuaries, bar-built estuaries, and estuaries produced by tectonic processes.

Drowned river valleys are perhaps the most familiar, occurring along most of the U.S. Gulf and Atlantic coasts. Because they are generally confined to coastlines with relatively wide coastal plains, these waterways have also been called coastal plain estuaries. These estuaries are wide-spread throughout the world and are often the sites of major ports. Philadelphia on Delaware Bay and London on the Thames River estuaries are two examples.

Bar-built estuaries, such as Pamlico Sound, are usually shallow and often too small to serve as ports for modern tankers. Fjord-like estuaries, such as Puget Sound, and tectonically formed estuaries, such as San Francisco Bay, are more acceptable for tanker ports.

In a typical estuary, the salinity of the water increases with depth as well as in the seaward direction. There is usually a surface layer in which the vertical salinity change is small, an intermediate layer in which the salinity increases rather rapidly with depth, and a deep layer in which the rate of increase of salinity with depth is small, as in the surface layer. Vertical diffusive and advective mixing takes place between the surface and bottom waters.

Tidal currents, ebb tides, and floods usually dominate in estuaries. Superimposed on the tidal currents is a net circulation pattern (known as the estuarine circulation) in which there is a net seaward flow in the surface waters and a net flow from the mouth toward the head of the estuary in the deeper waters. There is also a small net flow from the deeper layers to the surface layers. The volume of water flowing toward the head of the estuary (per unit time) decreases from the mouth to the head of the estuary, since water is simultaneously being moved upwards (entrained) from the deeper layers to the surface layers. Consequently, the amount of seaward flow of surface waters increases from the head toward the mouth of the estuary.

Being less dense than seawater, any crude oil or petroleum product spilled in an estuary moves with the surface waters in the net flow toward the mouth of the estuary. Mixing of the surface and subsurface waters leads to horizontal and vertical spreading. Soluble constituents will eventually be added to the deeper layers and have a net flow toward the head of the estuary. Floating oil is ultimately flushed from the estuary in the seaward flow of the surface layers or is washed ashore to be deposited on beaches or trapped in wetlands surrounding the estuary.

The basic estuarine circulation pattern prevails in most of the coastal ocean where the input of the freshwater from river runoff and rain exceeds the loss by evaporation from the surface. Thus, over the continental shelf, the surface waters generally have a net motion seaward, while the near bottom waters move generally toward the shore as well as moving along the coast.

Shores — Beaches and Wetlands

The shore is that strip of land bordering the ocean that is alternately exposed or covered by waves or by the changes in water level due to the tide. Where wave action is relatively vigorous and the supply of sand is adequate, the shore is marked by a beach — a dynamic deposit of loose sand or gravel. In more protected areas, wetlands (either marshes or swamps) border the shoreline. In those areas where the land is rising due to mountain building or the recent retreat of the continental glacier, the shore may be marked by a rocky coastline, such as the New England coast (north of Cape Cod) or parts of the Pacific coast of the U.S.

Onshore winds and wave-induced water movements move spilled oil toward the shore. Indeed, some of the more dramatic effects of spilled oil have been observed along the shore and in wetlands subjected to large spills of oil such as the oil from the Torrey Canyon along the Cornish coast of Great Britain or the Brittany coast of France.

Rocky shores are typical of steeply sloping coastal areas where the sea acts directly on resistant rocks. In many of the areas used for supertanker ports (especially fjord-like estuaries), rocky shorelines are common, owing to the recent glaciation that provided the deep waters nearshore suitable for the deep draft vessels and also removed soils and unconsolidated deposits from the coastal areas. In such areas, marine organisms have typically attached themselves to the rocks which support extensive communities of marine animals and plants.

Where sands and gravels are more abundant, beaches form under the action of the waves. Built of loose sand or gravel, the beaches build outward and retreat in response to changing wave conditions. Relatively few organisms grow in or on the beaches, but they are not wholly devoid of life. On the lower parts of the beach where wave action is less vigorous, burrowing animals can usually be found.

Wetlands are low-lying coastal areas that are intermittently covered by tidal waters during the normal spring tides (the highest tides that occur every two weeks). Also included in the wetlands are the tidal channels through which the waters flow into the wetlands during a flooding tide and drain out again during an ebbing tide. Tidal marshes are typically covered by marine plants — either algal mats or various kinds of marine grasses, depending on the location of the marsh.

Some wetlands are protected from direct wave action by the topography of the shoreline or by barrier beaches. However, the relatively fine-grained sand and mud deposits in the marsh areas are stabilized by the growth of the plants which tend to trap not only sediment, but also floating debris, including petroleum. Thus, these marshes may be subjected to oil accumulation when the spills occur well outside the marsh itself.

Wetlands are common along the Atlantic and Gulf coasts of the U.S. but also occur along the Pacific and Alaskan coasts as well. They are highly productive of plant material and thus are important sources of food for many marine organisms. Furthermore, they serve as home or nursery grounds for many species of marine organisms at some stage of their life cycle.

Wetlands are unlikely to serve directly as deepwater port areas; instead, they may be involved in pipeline construction from a port facility to onshore storage. Owing to their abundance along the Atlantic and Gulf coasts and the nature of water movements in them, wetlands are likely to be affected by oil spilled in the coastal waters or in harbors.

Man's Utilization of Coastal Zones

Man uses the resources of the coastal zone in many ways. Major uses fall into the following categories:

1. Waste disposal (municipal sewage, industrial wastes)
2. Shoreline development (industry, housing, parks, etc.)
3. Exploitation of living resources (fishing)
4. Recreation (swimming, boating, sport fishing, etc.)

5. Water resources (municipal and industrial supplies)
6. Transportation (shipping, waterways, harbors)
7. Exploitation of nonliving resources (oil, gas, gravel, sand, etc.)

Engineering works have been developed to alter the coastline to suit man's desires. However, imperfect knowledge of the nature of the forces acting in the coastal zone and their effects has resulted in unexpected or undesirable side effects. Frequently such alterations create substantial conflicts between uses. For instance, use of the waters for waste disposal, resulting in pollution, directly conflicts with recreational and fishing uses. The filling of wetlands for use in residential, industrial, or port construction conflicts with wildlife and estuarine preservation, recreational uses, and fishing interests. Dredging of ship channels for navigation can alter salinity, temperature, and numerous other environmental parameters which conflict with almost all uses of the coastal zone area.

GENERAL DESCRIPTION OF PIPELINE CONSTRUCTION AND OPERATIONS

The main interest of this report is an evaluation of pipeline construction and operations on the beaches and in wetlands area. It is important, however, to understand some of the offshore construction methods and considerations in order to appreciate the pipeline characteristics and conditions as it reaches the shore area. It is conceivable that some of these methods will be used in the open waters of estuaries as well.

Offshore Methods and Techniques

Pipelines are generally thought to be among the safest means of transporting liquid products. It is highly unlikely that any form of transportation other than pipeline will be used to move oil from the deepwater port to the mainland.

Construction and Burial

Pipelines laid offshore are constructed and laid by several different methods, depending mainly on the size, location, intended use, and cost. One method, pipepulling, involves the use of barges and tugs to pull sections of welded pipe from an onshore launchway over the preselected right of way. These sections may either be dragged along the bottom or suspended by floats. There are at last three limitations to this system. First, an extensive section of shoreline, roughly perpendicular to the shore, must be available for the fabrication and launchway site. (Alternately, it is possible although more costly to use a launching jetty constructed from the beach out over the water.) Second, the total length of pipeline that can be laid is limited. One company estimates the limit to be 100,000 ft for smaller-diameter pipe. Third, the pipeline right-of-way must be essentially a straight line.

The second method, used in nearly all cases for large-diameter pipelines, involves the welding together of short sections of pipe on a barge while simultaneously moving the barge forward and allowing the completed expanse of pipeline to sag downward and lay on the seafloor. This operation begins at the offshore location and proceeds toward the intended onshore terminal. The advantage of this system is that the pipeline right-of-way need not be straight and that any diameter of pipeline can be laid in this way. The main disadvantage is slow rate at which the laying proceeds; average rates are about 300 to 800 ft/hr.

The third method has become increasingly popular in the last decade for laying smaller-diameter pipelines and involves the use of a lay-barge equipped with a large reel or spool of coiled pipe. With the reel-pipelaying technique, miles of pipe are welded together onshore and the appropriate coatings are applied. Then the pipe is wound onto a large-diameter reel which is mounted on a barge or other floating vessel. The vessel is then transported to the construction site. As the barge is towed along the right of way, the pipe is pulled off the reel through straightening equipment, and continuous lengths of pipeline are laid on the seafloor. The reel method has several advantages. With no welding and little crane work being done on the barge, the operation is much less susceptible to interruption by bad weather and high seas. A thicker-walled pipe can be used, eliminating the necessity of a concrete coating for negative buoyancy, increasing the pressure rating, and adding significant corrosion allowance. This method allows pipelaying to proceed at a much faster rate than other methods — up to 5,000 to 10,000 ft/hr.

The reel method has two principal disadvantages. First, economic considerations limit the diameter of the pipe that can be laid to 12 in. Second, during pipehandling operations, pipe coatings are subject to occasional damage, necessitating repairs and thus slowing the rate of pipelaying.

Pipelaying capabilitics have now been extended to the point that 12- to 18-in. diameter lines can be installed in water depths exceeding 1000 ft. Connecting pipelines to seafloor facilities in water depth beyond diver effectiveness (about 1000 ft) will require remotely operated pipeline connectors. Some of these are now in various stages of development.

In depths under 200 ft, present OCS administrative procedures require burial of the pipeline. Burial is effected by jetting sediment away from underneath the pipeline and allowing it to sink into the resulting trench. The equipment used in this operation consists of a work barge equipped with high-volume/high-pressure water pumps and air compressors. From the barge, a multiple-membered towline consisting of a strength member, water line, and air line extends downward to a U-shaped structure which saddles the pipeline and glides along it on rollers. Affixed to the U-shaped jetting device are several nozzles which direct water and air, under high pressure, ahead and below the pipeline. Sediments are blasted out of a narrow trench by the water jets, partially lifted by the air and deflected to the sides by various types of fins. The suspended sediments falls diffusely along either side of the trench. As the jetting device is pulled forward, the pipeline settles into the trench and is partially buried quite soon by the reworked sediment as it slips and settles back into the depression. Complete burial and restoration of original bottom contours may require additional time. In shallow waters, experience has shown that contour restoration to be quite rapid, whereas in deeper waters, more than a year may be required.

Even though a buried line is protected from fluid forces, it is not necessarily stable. If it is too light, it will gradually work its way up through the soil and become exposed to the water forces. If it is too heavy, it will gradually sink in the soil and impose additional tensile stress in the line. Design procedures for determining the vertical stability of the line in sands and clays have been developed and are available in the industry.

Difficulties have been experienced in burying pipe in cohesionless sands. In this case, the sand will often refill the jetted trench before the pipe can settle into it. Another method, fluidization of the sand, enables successful burial in this type of substrate.

In waters beyond the 200-ft contour, pipelines are not buried. Industry spokesmen maintain that as yet, burial in waters deeper than 200 ft is not economically feasible, but that new equipment could be available in the future.

To prevent corrosion, pipelines are carefully coated with such materials as epoxy compounds to thick asphaltic mastic. If extra weight or mechanical protection is needed, these, in turn, are covered with a layer of dense concrete. The lines are protected from electrolysis by both impressed-current systems and by sacrificial anodes (zinc is commonly used). Corrosion prevention measures are now required by 49 CFR Part 195. Although offshore pipelines are relatively inaccessible as compared to onshore pipelines, they nonetheless can be repaired by divers. Experimental dives have been made to 1000 ft, but work at this depth is difficult and expensive. Methods of using submersibles to latch on to a subsea line and repair it with mechanical arms and special tools are under study and nearing the point of practical demonstration.

As in the case of workover operations, the expense of the pipeline installations, coupled with the catastrophic implications for the local marine environment should a major break occur, have combined to dictate a highly conservative design, emplacement, and operating philosophy.

As the pipeline construction approaches and traverses the shoreline, it is buried deep enough to avoid its being exposed by storm-associated beach erosion. From this point, the pipeline construction will be extended toward a storage facility, wharf facility, or a major existing pipeline system, in turn leading to a processing facility, refinery, or interstate oil line.

Pipelines — Operation and Maintenance

The safe operation and maintenance of a pipeline system requires several redundant monitoring systems to ensure the integrity of the line and detect leaks. The primary leak detection system in use (required on all lines built after March 31, 1970 by 49 CFR Part 195.404 and 195.408) is a set of automatic pressure-sensing recorders on both ends of each pipeline system. These recorders are equipped with a built in alarm system which either shuts down the flow automatically or sounds an alarm to alert personnel of an abnormal pressure level. In this way, a leak of substantial rate is detected immediately. This system is insensitive to very small leaks which do not produce a decrease in line pressure.

The second system of leak detection is the routine patrolling of the offshore and wetlands routes by boat or aircraft, and onshore by wheeled vehicle or aircraft. This type of monitoring would result in the detection of all sizes of leaks, of course, but would be of little consequence in preventing the loss of a large amount of petroleum in the event a large line were severed. The appeal of a system of regular pipeline patrolling is that it allows detection of small leaks and therefore compliments the pressure-sensing system described above.

The third system for leak detection consists of a series of volume-recording flow meters on either end of a pipeline system. Because nearly all crude oil moves from OCS areas to shore by common carrier lines, it must be metered in the offshore pipeline gathering system and again at the onshore pipeline terminal in order that each producer be properly credited for his share of the common stream. This flow-monitoring system has been designed so that the flow sensors continually indicate net input and output in real time, so that attendant personnel are able to discover a decrease in output and alert appropriate stations of the possibility of a leak.

One more safety feature which would be built into all pipelines resulting from this proposal, according to industry spokesmen, is that remotely operated mainline block valves will be provided remotely controlled pipeline facilities in order to allow isolation of segments of the pipeline.

Pipeline Size

The size of pipeline from the deepwater port facility to the onshore storage will depend on the proposed crude oil throughout, the number of berths for each complex, and the distance the oil is to be moved. At present, a 48-in.-diameter line is the largest submarine pipeline capable of being installed by lay-barges. However, in cases where terminals are more than 10 mi offshore, the use of 54- or 60-in.-diameter pipelines would greatly reduce the required pumping horsepower. These sizes may become economical as the state of the arts improves and lay-barges become available to handle these very large submarine pipelines.

It is expected that pipelines from the first deepwater ports will range between 36 and 48 in., and they will be buried in concrete. Existing administrative regulations on the OCS require that pipelines be buried if the water depth is less than 200 ft.

Inshore Methods and Techniques

The location of each wetland crossing should be carefully selected to minimize impact on the environment. Several pipes that can be routed in one canal (in place of separate canals for each line) should be utilized wherever possible. The route selected should be carefully surveyed so that all information will be available for the design. The pipe should be laid so that it is secure from currents, tides, wave action, and floods and will not be adversely affected by channel changes or scour.

General Methodology

The method used in installing the crossings will generally be by dredging a canal with a barge-mounted dragline or clamshell bucket especially designed for this work. The bucket is located on the front of the barge, and the width of the canal is controlled by the width of the barge, usually 30 to 40 ft. The spoil will usually be placed along the sides of the canal, and the bottom of the canal will be excavated so that the top of the pipe will be just below the bottom surface. Following the excavation, a pipelaying rig is used to lay the pipe in the excavated trench. Backfilling is not normally required, with the only cover being from normal siltation. From an engineering standpoint, routing the pipelines through canals is preferred, as it does not present an obstruction and the pipeline is readily accessible. Helicopter patrols can easily spot leaks, repairs are simplified, and inspection of the pipe for corrosion and adequacy of cathodic protection can be done quickly and easily.

Weighting

Extra weight must be added to the pipe so that it will not float when empty. Extra-heavy pipe may be used for all or part of the additional weight, concrete weights may be used, or a continuous sheath coat of reinforced concrete may be used. Negative buoyancy of 65 to 100 lb per lineal foot is required where pipes are subject to tidal action, wave action, or floods, but a negative buoyancy of as low as 10 to 25 lb per lineal foot can be used in still water. Each installation must be studied with regard to the hydraulic conditions and the specific gravity of the material surrounding the pipe. For example, the specific gravity of water-saturated silt or saltwater must be considered.

Pipeline Location

The hydraulic conditions must be studied carefully. Where canals cross areas where water is flowing, closure structures are often required to retain as closely as possible the natural condition of the wetlands. Foundation conditions and bank stability must be examined closely to prevent sag or overbending which can overstress the pipeline. Special regulations applicable to crossing of navigable waterways must be adhered to. Special markings, valving, and other similar requirements are typical. The depth of burial will also be controlled by present or future navigation channel depths and importance of the waterway. For minor waterways, depth of burial required may be as little as 6 ft, whereas across major waterways, the depth of burial required may be 25 ft or more below the channel bottom to make certain the pipes will not be damaged by ships anchors. Usually pipelines crossing levees will have to be installed over the levee and covered so that no excavation of the levee will be necessary. Pipelines crossing sand pits and beaches between the ocean and the wetlands must be buried to a depth sufficient to ensure that storms and extreme tides will not scour out the sand exposing the pipeline.

Corrosion

Corrosion protection of the pipeline will be required, and a special corrosion study should be made to determine the most suitable combination of coating, cathodic protection, and extra pipe thickness as a corrosion allowance. Coating and cathodic protection will usually be required. Extra wall thickness and reinforced concrete coating will give protection from corrosion and also additional weight and strength. In brackish and saltwater environments, marine life can often be so destructive to steel and bituminous coatings as to require sheathing the line with concrete. The coating must be sufficiently strong to resist damage from handling, laying, backfilling, and scour effects of the sand or soil. A smooth, reinforced concrete coating will provide this protection and is preferred. Cathodic protection should be provided for the lines and provisions made to periodically test the effectiveness of the system after installation.

Welding

Welding procedures and qualification of welders should be in accordance with the applicable standards, and in addition, for sections crossing wetlands or waterways, all girth welds should be 100% radiographically or ultrasonically tested.

Alternatives

Alternatives to the methods of installation described may be advantageous in a few instances. Overhead crossing in some cases may be desirable. The pipeline can be supported by simple "A" frames, suspension cables, or by a structural bridge. In installations of this type, special studies are required of foundation conditions, loadings including earthquake, wind, and ice, permissible spans and pipe thicknesses required, and clearances required considering high water. Alternative methods of laying the pipe which may be desirable in a few instances are dewatering and dry excavating, using a sand pump to remove sand or silt from underneath a line to lower it into place, or using equipment designed specially to hydraulically jet out a trench under water.

Description of Pipeline Construction in Marshy Areas

Sometimes pipelines transverse many miles of marshes and bays or bayous before reaching ground firm enough for conventional pipe laying. Two methods are used for all medium- to large-diameter pipe. These are commonly referred to as "push" or "shove" and "flotation", and a pipeline may employ both, depending upon the nature of the marsh terrain traversed. These lands vary widely in nature from soft to firm, and this significantly affects the method of pipe laying. Surveillance and maintenance considerations may also affect local application of alternative methods.

"Push" Method of Pipelining

The "push" or "shove" technique is possible only where the marsh is firm. In this technique, a narrow, relatively shallow ditch is excavated by a dragline or clamshell digger from the bank. By using a marsh buggy base or by using runners or pads to spread the weight, the damage to the bank is minimized. The ditch may be 4 to 6 ft deep by 8 to 10 ft wide. The pipe sections are joined together at the point of origin of the ditch, and the line is given temporary buoyancy by strapped floats and pushed or shoved down the ditch. A section as much as 15 mi long can be installed in this fashion. After being floated into place, the floats are cut loose, and the line is allowed to sink to the bottom of the ditch. Typically, there will be approximately 4 ft of water above the pipe. The ditch may be left open, but is more frequently backfilled. Even with firm marsh, there is generally sufficient subsidence and shrinkage that the spoils will not completely fill the ditch. However, there usually is no canal after completion.

The shove technique is less costly than using flotation barges and is preferred where possible.

"Flotation" Method of Pipelining

The second method of pipe laying utilizes a flotation canal to provide access for the pipe laying equipment. Such a canal may be 40 to 50 ft wide and 6 to 8 ft deep and may have an additional trench in the bottom to provide 10 to 12 ft clearance above the top of the pipeline.

The pipeline is constructed on a series of lay barges and passed over the stern of the train. The pipe is large and heavy, and massive equipment is needed to manipulate it. For example, a standard 40-ft section of 36-in.-diameter pipe weighs approximately 8000 lb. After the addition of a corrosion coating and 3 to 4 in. of concrete to give it negative buoyancy, a 40-ft section weighs about 34,000 lb. Equipment to handle weights of this magnitude cannot be supported by the marsh.

This type of canal is excavated by a flotation dredge which normally piles the spoils upon each side to form a low levee. Characteristically, where this type of canaling is carried on, the marshes are soft and unstable, sometimes to the point of being near-floating marshes.

Generally, the spoils are piled back some distance from the canal to leave a 30- to 40-ft berm between the canal and the levee. The levees are characteristically low and flat. Depending upon the width and depth of the canal (which determines the quantity of the spoils) and on their stability, a levee may be 3 to 5 ft high and possess a base width of 50 to 85 ft. The high water content (50 to 85%) and the organic nature of the excavated muck lead to major shrinkage and subsidence when piled on top of marshland with similar properties. Height reductions of 50% are possible. The paramount factor in the permanence of a levee is the type of marsh upon which the levee is placed. The condition of the marsh at the time the levee is placed is of great importance. When water is over the marsh, the surface material is supersaturated and becomes very weak and almost fluid. Poorly shaped levees which spread across the marsh result from this condition.

Because of these factors, there are never enough spoils from the excavation to backfill a flotation canal. Where canals traverse state-owned land or wildlife refuges, very stringent conditions may be attached to the right-of-way grant, with the objective of minimizing impact on the territory, and these may include backfilling a flotation canal. In one recent example, nearly 2 mi of flotation canal crossing a wildlife refuge were backfilled. However, as usual there was insufficient spoil material, and additional backfill (nearly 160,000 yd^3) had to be dredged from a nearby bayou and lake. Backfilling with foreign material on a large scale is probably economically unfeasible, and the environmental impacts of the extensive additional dredging required are unknown, but probably not insignificant.

In the past, levees were characteristically continuous, with few or no breaks. More recently, the Wildlife and Fisheries organizations have required openings to be cut in levees in order to minimize disturbances to existing drainage and use patterns.

In the course of laying such a line through marshland, numerous bodies of open water will also be traversed. The same equipment may be used, although only a trench will be dredged, or the assembled line can be jetted into place. Hydraulic dredges are also used for open-water traverses. Hydraulic dredge spoils may be pumped to nearby land, dispersed over the nearby area, or piled up in a spoil island, depending on the particular situation.

Treatment of completed canals, whether "push" or "flotation", is a matter of negotiation between the pipeline owner and the owner of the land being traversed. Land owners may require bulkheads or plugs or dams wherever a canal intersects another waterway, in order to minimize erosion and to prevent navigation traffic which is a prime cause of erosion. There is a real economic incentive for the landowner to insist on minimizing the opportunities for erosion.

As an example of the preventive measures taken, for one 284-mi line constructed in the late 1950s which crossed 135 navigable streams, 240 bulkheads were constructed.

The types of dredging equipment used and the various methods of spoil deposition employed are shown in Table 1.

Construction Operations

As with any major engineering effort, there are a number of actions and effects from the general pipeline construction operations irrespective of the type of canal constructed. Either type requires surveying and an alignment established and marked. Marsh buggies have normally been used for this operation; these may have a permanent effect on the marshes, especially some softer ones. Even though wide tracks are used, and the unit pressure on the soil is low, the weak marsh structure is compressed, and depressed tracks are left. These may not be self-repairing in some cases because of their depth and may act as erosion foci.

Similarly, when bank-supported draglines are used to dredge push canals, the berm may be damaged, even when pads are used, if the marsh is soft.

In the construction of a flotation canal and laying of a pipeline, there also can be erosion effects upon the canal from the ancillary boat traffic bringing men and materials to the site but this is a short-term effect.

Evaluation of Incidents of Pipeline Spills

Pipeline spills evaluated include casualties as well as operational failures associated with the underwater pipelines that could carry oil from deepwater port terminals to refineries or storage areas onshore.

Information from the U.S. Geological Survey (USGS) reveals that between 1967 and 1972 there were an average of 975 mi of underwater pipelines carrying oil under their jurisdiction. During the 5-year period, a total of 1.48 billion barrels of transported oil was achieved in the network for an average of 0.8 million barrels of oil per day. During the period, a total of 10 spillage incidents occurred with a total outflow of 175,748 barrels (see Table 2).

These figures work out to a rate of 2.05×10^{-3} incidents per mile-year. The 10 incidents had an average spill of 175,748/10 or 17,575 barrels per incident.

Using this information, it may be projected that a 10-mi underwater pipeline transporting oil from a deepwater port termimal to a facility onshore, or 2.05×10^{-3} incidents per mile-year × 10 mi × 17,575 barrels per incident = 360 barrels per year throughout.

Because the pipeline from the port site to the shore would be buried (administrative regulations now require that pipelines be buried if the water depth is less than 200 ft), the possibility of an anchor dragging-caused spill might well be eliminated. Excluding, therefore, the 160,000- and the 6000-barrel spill from the data base, results would be 9748/10 = 975 barrels per incident or 2.05×10^{-3} incidents per mile-year × 10 mi × 975 barrels per incident = 19 barrels per year, or something less than 0.05 barrel per day for a 1 million barrel per day throughout.

The analysis indicates that the single most cost effective method of reducing total oil spilled is by using pipeline — instead of vessels — to transship oil from deepwater ports to shore. As a result, the risk of damage to the environment would be significantly reduced.

Table 1
DREDGING EQUIPMENT AND SPOIL DISPOSITION

Dredging equipment and method	Principal use	Disposition of spoil	Spoil control	Comments
Mechanical Dredge				
Bucket and dragline	Small channels	Place adjacent to channel	Small spoil area; minimum turbidity	Usually the preferred method; problem exists of blocking drainage and reducing water exchange or circulation
		or Place on barge	Remove from site	High cost usually prohibitive
Hydraulic Dredge				
Hopper	Large channels	Remove from site to designated spoil area of preferably low-value habitat	Excellent at site; moderate at discharge point where turbidity is frequently high	Used where spoil area is not available near site or where adjacent spoil deposition would increase maintenance dredging
Pipe discharge	Large and medium-size channels	Place in mounds adjacent to channel	Very poor unless spoil is retained by ring levee; turbidity is high without ring levee and spoil area is large	Danger exists of segmenting bays and blocking drainage, tidal exchange, or circulation; destruction of shallow bay and marshes by filling creates a major problem
Side broadcast discharge	Large channels	Scattered in layer over large area of bottom adjacent to channel	Depth of spoil on bottom can be controlled; spoil area is large, turbidity is high, and silt is dispersed by currents	Can prevent bay segmentation, but very large areas of water bottoms can be silted; needs more study

From Chapman, C., Marsh and Estuary Management Symposium, Louisiana State University, Baton Rouge, 1972. With permission.

Table 2
UNDERWATER PIPELINE SPILL DATA

Date of casualty	Volume spilled (bbl)	Cause of spill	Location
02/07/67	65	Unknown	Offshore Louisiana
10/15/67	160,639	Anchor dragging	Offshore Louisiana
03/12/68	6,000	Anchor dragging	Offshore Louisiana
01/24/69	100	Pipe cleaning operation	Offshore Louisiana
02/10/69	342	Unknown	Offshore Louisiana
02/11/69	7,532	Unknown	Offshore Louisiana
08/10/69	50	Unknown	Offshore Louisiana
12/16/69	900	Unknown	California
12/31/70	50	Unknown	Offshore Louisiana
11/14/71	70	Unknown	Offshore Louisiana
Total	175,748		

From The U.S. Geological Survey, Washington, D.C., 1973.

GENERAL CHARACTERISTICS OF THE MARSHLAND ECOSYSTEMS

Terrestrial Ecology

Land development in estuaries may proceed by organic accretion as well as by siltation. The mechanisms of this development will be considered. The various plant communities within the marshes are not equally productive with respect to organic build-up, just as they are not equally productive for support of differing shellfish or waterfowl species.

Along the seacoast or sea rim, the immediate action of the tides produces an area devoid of vegetation build-up or peat formation. Here, wave action piles up layers of sand and shale (marine sediments) on top of the river sediments. The sea rim is characterized by only pioneer species of plant life and an abundance of shore birds.

Behind the protective sea rims are vast areas of salt marsh. Salt marsh may also extend directly into adjacent open seawater areas protected from wave action. This occurs particularly in the numerous bays and terminal ends of the major stream channels. Frequent inundation by saltwater as a result of natural storm tides is necessary to maintain usual salinities. Protection against the abrasive action of waves is afforded by the sea rims and bays and the natural levees along the transecting waterways. Organic accretion in these areas is moderate. The high production levels characteristic of smooth cordgrass (*Spartina alterniflora*) would provide for extensive accretion if it were not for the abrasive action of the tides. The vegetation produced in these areas is mechanically broken down, flushed by the tides, and enters the detrital food chains. It is estimated that as much as 50% of the *Spartina* organic matter is flushed from the estuary and that 80% of that amount finally gets offshore. This in part accounts for the enormous value of this marsh type in shellfish and fish fry production.

Further inland where the pressures of the freshwater flow compensate for tidal pressures, a mixing zone occurs. These brackish marshes are the site of one of the most productive areas in the world. Vegetative land accretion in terms of peat accumulation can be very dramatic in the brackish marshes. The high nutrient levels provided by the freshwater flow, coupled with the preserving qualities of the saltwater, allow the accumulation of many yards of peat in the more productive areas. The most productive areas of coastal marsh are intermediate zones where salinity is not too high to restrict growth, and yet available nutrients are in high enough concentrations to allow good growth. These moderately saline marshes are in the greatest danger of being destroyed by saltwater intrusion and industrial activities.

In the areas normally protected from saltwater, a different marsh type develops. While plant species differ, a good nutrient supply makes productivity nearly equal to that of the brackish marshes. Peat formation is, however, somewhat less in these areas due to the lack of protection against decay that the higher salt content of the brackish and saline marshes provide. Temporarily increased salinity from storm tide action, if not sustained, will usually be flushed from the system before serious changes occur. With prolonged salinity increases, existing salt-intolerant species will die out, followed by the growth of more salt-tolerant species until a new salinity equilibrium is established. The lag time in reestablishment of an equilibrium depends on the source of the initial salinity change and its duration. Aside from the initial impacts of such losses of vegetation, large expanses of mudflats may result if salinities do not become restabilized.

Coupled with the varying potential of the marshes in organic detrital production, substantially different substrate stabilities occur. The stability of a particular site will greatly influence the impact of construction on that area. A continuum of substrate makeup exists from the pure silt soils to the pure organic soils. Both of the endpoints of this continuum represent highly unstable soils. The uniformity of both the silt and muck soil provides for little consolidation and thus little stability. However, in areas where vegetation products and silt have mixed, soils may have greater stability. Peat areas and others, where root and tuber production are extensive, are characteristically very stable. It is these areas that represent the so-called "good" spoil of pipeline construction.

The marshlands have a wide range of environmental characteristics important to many forms of terrestrial and marine life. High productivity of fur and waterfowl, oysters, fish, and shrimp indicates the magnitude of importance to man.

Plant Communities

Sea Rims

The so-called sea rims develop as a result of wave and tidal action along the shoreline in areas where there is limited natural subsidence and an availability of sand and/or fragmented shells. High salinities, numerous shore birds, and limited pioneer vegetation characterize this habitat. Similar "beach" areas developing in a bay not contiguous with marsh or land are called "shell shoals".

Salt Marsh

The saline vegetative type makes up about 22% of the coastal marshes. The saline marshes are represented by nearly pure stands of the smooth cordgrass (*S. alterniflora*). Salt grass (*Distichlis spicata*) and black rush (*Juncus roemerianus*) are present though much less abundant. The profile of this community is very short, 2 to 4 ft. The deeper lagoons are usually free of standing vegetation, but are characteristically surrounded by pure stands of *S. alterniflora*. The upper few feet of marsh are well consolidated by the roots and rhizomes of the vegetation. Below this, the substrate is almost totally unconsolidated. These subsoils are almost pure silts or silt and clay and contain very little organic matter.

Brackish Marsh

The flora and fauna of the brackish or intermediate salinity indirectly reflect this salinity. While most of these organisms are adapted to a fairly wide range of salinities, extended periods of inundation with freshwater or seawater would change and eliminate some of them. Organic content may be extremely high. Deep layers of peat are not unusual.

The brackish marsh contains a number of species of vascular plants, with marsh hay cordgrass (*Spartina patens*), the dominant species (as much as 55%). The brackish vegetative type makes up about 30% of the marsh area. The brackish marshes are most often characterized by an association of *S. patens*, salt grass (*D. spicata*), and black rush (*J. roemerianus*). This vegetation is taller than the salt marshes, reaching heights of 4 to 6 ft. While this marsh type can and often does stand uniform for many miles, significant variations can occur with slight changes in elevation or salt content. One of the major variations is the cane zone (*Phragmites communis-Spartina cynosuroides*). Quill cane and roseau cane are major components of this taller plant association. Heights of 6 to 8 ft are not uncommon. The water table in these cane zones is usually below the surface. Another major variation in the brackish marsh is the three-cornered grass (*Scirpus olneyi*) marsh. While this species is often present in many brackish marshes, it frequently forms a dense uniform stand of large area. This is a subclimax stage of the *Spartina-Distichlis-Juncus* association and will be naturally displaced if undisturbed. This three-cornered grass or *Scirpus* marsh is a favored habitat for the muskrats and waterfowl, and as such is the site of extensive resource management.

Intermediate Marsh

The intermediate marsh vegetation type comprises only about 16% of the marsh area. *S. patens*, comprising as much as 34% of the vegetation, remains dominant in these marsh areas. Roseau cane (*P. communis*) and bulltongue (*Sagittaria falcata*) are also major species in these areas.

Fresh Marsh

A diversity of associations and a larger number of species are found in the fresh marsh. The fresh marsh area in southern regions are almost completely dominated by alligatorgrass (*Alternanthera philoxeroides*) and waterhyacinth (*Eichornia crassipes*). These two exotics are so well adapted as to have displaced the natural vegetation in this area. Other widespread plant associations of the fresh marshes are edaphically controlled by water level and water quality for the most part. Some of these include large expanses of maiden cane (*Panicum hemitomon*). Spikerush (*Eleocharis* sp.) and bulltongue are also major components. Other species present are cattail (*Typha*sp.), bulrush (*Scirpus californicus*), saw grass (*Cladium jamicense*), and wapato (*Sagittaria platyphylla*), with alligatorgrass and waterhyacinth. These marshes may be attached or floating. Floating marshes are characteristic of some areas of fresh and brackish marshes. They may occur anywhere the clay pan has subsided, allowing the root consolidated vegetation mat to float on a sea of water and organic muck.

Island Beaches

The dynamic nature of the coastline is shown by the development of stranded beaches within the marshlands. Earlier deltaic formations pounded by storm tides formed beach lines that have now become stranded inland. These terrestrial "islands" are usually parallel to the ocean. These islands differ from the sea rim because of its terrestrial sediment composition rather than the marine origin of sea rim sediment. Many of these islands have sunk below mean water level and now provide an area of sandy bottom and an important substrate for oysters and other shell creatures; however, others remain above water level, producing high ground, possibly 2 to 3 ft above normal high tide. This condition provides for the occurrence of a different habitat. Similar species are seen to invade the natural levees of streams and rivers and the spoil piles along canals — areas in which the sand content of the soils may be elevated over the surrounding marshes. While absolute species composition is regulated by substrate composition (sand and silt content) and water level, certain plants are characteristic of the island and levee areas. The lower areas are occupied by various grass and sedge species, depending on the salinity and water level.

Swamps

In certain areas adjacent to higher ground in the marshes and in protected inland areas, successional patterns have led to various tree associations. Two basic types are defined consisting of saltwater and freshwater swamps.

Mangrove swamps occur only in southern waters. They are composed almost entirely of the shrubby honey or black mangrove (*Avicennia nitida*) occupying high saline environments. The crest of a "ridge" in the salt marshes may provide a habitat for this species. As little as 2 in. above the adjacent salt marsh, this zone is usually above high water level. Very high soil salinities are characteristic. Reaching a height of approximately 25 ft, the honey mangrove is an evergreen shrub. Aerial roots or "knees" are characteristic physical features of the plants. With these structures, the mangrove has the potential to evade more deeply into the marsh, but is usually restrained from doing so by frequent marsh fires. The mangrove is also intolerant of cold and is killed back by any frost or freeze, further reducing its potential for spreading.

The cypress-gum swamps of the freshwater and nearly freshwaters are characteristic of extensive areas and scattered areas within the open freshwater marsh where water conditions are static and where some protection from storms and fires is afforded. Bald cypress (*Taxodium distichum*) and tupelo gum (*Nyssa aquatica*) are codominants in southern areas. Other tree species include water ash (*Fraxinus profunda*) and swamp maple (*Acer rubrum*) which tend to dominate as one moves in a northerly direction. Where the swamp is open, numerous herbaceous species prevail. Along the swamp borders are a great number of black willows (*Salix nigra*) and buttonbush (*Cephalanthus occidentalis*). Other species associated in areas where inundation is only periodic include sweetgum (*Liquidambar styraciflua*), water oak (*Quercus nigra*), hackberry (*Celtis laevigata*), cottonwood (*Populus deltoides*), sycamore (*Platanus occidentalis*), and box elder (*Acer negundo*).

Animal Communities

Stratification within a community often plays an important role in the number of kinds of animal organisms that community is able to support. The hardwood forest communities of the swamps and islands do provide multiple strata for support of types of organisms found nowhere else in the marsh, e.g., in the water, in emergent herbaceous vegetation, or in shrubs. While the typical marsh plants and animals are found associated on the ground level within these communities, even here new species will occur when shading or water level changes bring about conditions that are different from the adjacent swamp. In effect there may be three or more different levels supporting corresponding plant and animal associations within one community type. The mangrove swamps (*Avicennia*) promote a similar effect. Here, however, fewer strata occur, and the single species dominance of the mangrove reduces the available number of ecological niches. Still, there is one association of plants and animals associated with the root zone and another associated with the canopy of the mangrove swamps.

Furbearers

Species hunted for their fur pelts include muskrat, mink, otter, bear, nutria, and raccoon. Other furbearers include opossum, skunk, lynx, fox, and beaver. The large numbers represented are a result of intensive management of large portions of our wetlands, particularly in the brackish marshes. For example, three-cornered grass marshes produce most of the muskrats and many of the nutria. To ensure good three-cornered grass production, extensive late fall burning of the marshes is practiced annually.

Other Mammals

The white-tailed deer and black bear are the only large mammals in the wetlands. The deer presently have a strong healthy population. They are abundant in all parts of the marsh except the saline where only a limited population exists.

While the bear is not officially listed by the USDI as an endangered species, its continuing presence in the southeastern U.S. will depend on a concentrated effort by man to protect the remaining habitat and in fact to enhance them. The effects of gas pipelining on the black bear have not been studied. However, no major problems are expected to occur normally. Difficulty would arise when water regimes for large areas were changed, followed by the loss of large areas of woodland.

Other important mammals include the fox, cottontail and marsh rabbits, and squirrels. All seem to have healthy populations and are surviving the effects of man on the wetlands because of their adaptability in habitat preference.

Insects

The mosquito is the major insect of importance to man in the wetlands, or perhaps, it is better to say the control of the mosquito is of importance to man. Large-scale chemical control is now limited mainly to the urban areas. Drainage of freshwater marshes is another important control method used in the suburban and urban areas. State authorities are now encouraging flooding as an effective and of less harm as a method of mosquito control. Flooding of an area by means of dams, levees, and weirs prevents the female from laying eggs and thus produces a significant population reduction. This procedure has also proven very beneficial to fish and waterfowl production as well.

Major carnivorous insects in the marsh include the praying mantis and dragon fly. Herbivores include, among many others, grasshopper species that graze on the grasses of the marsh. The importance of the grazing herbivores in the marshlands is less important as compared to their usual value within temperate ecosystems because of the particular energy dynamics of the marshes. The marsh system results in most of the grass biomass being broken down mainly by physical processes to small fragments where it is acted upon by bacteria and detritus feeders. The larger carnivores feed from this aquatic chain rather than from the terrestrial chain as would be the case further inland or on higher dry ground.

Reptiles

Snakes present in the wetlands include several species of water snakes, blue runners, ribbon snakes, black snakes, and bull snakes, as well as water moccassins, rattlesnakes, copperheads, and coral snakes. Information is needed on the potential effects of pipelining on the abundance and mobility of poisonous snakes.

Several species of freshwater and saltwater turtles are found in the wetlands. Sufficient quantities are found to make commercial harvest profitable in some areas.

The alligator deserves special reference, since it is officially listed as an endangered species. Through intensive study of alligator breeding habits and landcarrying capacity, officials have successfully returned a million acres of marshland to near capacity or better, and limited hunting has been allowed in some areas. Much work still remains to be done with the alligator.

Amphibians

Frogs are extremely abundant throughout the marshlands. The Wildlife and Fish

Commission has recently abandoned an investigation of bullfrog farming for scientific markets. A number of salamander species are also found in the marsh.

Avifauna and Its Habitat

There are hundreds of species of birds frequenting the wetlands, but rather than dealing with species per se, broad categories will be considered except where special attention needs to be drawn to certain problem areas of species. Under the various habitat groups, the following avian statuses, where salient or applicable, are considered: permanent, summer, and winter residents and migrants.

Permanent Residents

Grebes form a minor part of the avifauna and are fish feeders. They are, for the most part, dependent on small fish, although invertebrates are also eaten. The few hawks and owls that frequent the marshes on a permanent basis are mostly adaptable species and are primary, secondary, or tertiary consumers. One species that may be encountered there that should receive special consideration is the swallow-tailed kite. Kites in general seem to be seriously declining in parts of the southern U.S., and a couple have a tenuous hold on their environment or, in the case of the swallow-tailed kite, have an undetermined status. Some kites seemingly respond markedly to only minor habitat disruptions.

Pigeons, kingfishers, woodpeckers, or perching birds that may occur in this habitat are generally numerically healthy and widespread.

Four groups of major importance, in part because of economic reasons, are the herons, pelican-like birds, ducks and geese, and rails and their relatives. Ducks and geese are of major importance as wintering or migrant birds and will be considered under that heading. Rails and coots are prime inhabitants of this environment and of economic importance as a sport and food bird. One would *a priori* suggest that coots are versatile enough to be little affected by habitat changes, but rails demand special attention as they seemingly have a much more narrow range of tolerance and more specific habitat requirements. Both are rather general in food habitats although rails are primarily invertebrates, a fact which dictates within limits the habitat that can be used. Habitat considerations are principally concerned with water depth or density of vegetation.

The brown pelican has virtually disappeared from some states. The reason for this decline is apparently variable, but probably not related to habitat destruction. The question of habitat of proper quality and quantity becomes germane. Successful reproduction seemingly depends on a certain amount of seclusion for nesting and a proper spectrum of water depth from which to secure food.

Herons, egrets, and relatives are generally colonial breeders. The colonies are local in nature. That the habitat is not used with generally uniform frequency suggests the spotty distribution of limiting factors and perhaps semirigid requirements. However, many of the requirements may be a function of social convention, and the physical destruction of habitat holding colonies may be simply temporary. With regard to specific pipeline routes, this group of birds requires attention, as they form a conspicuous part of the environment both in feeding and nesting habitat requirements. They cannot use water over a certain depth as dictated by the length of bill or leg of the species involved. They may conflict with man's interests in the commercially important invertebrates should the habitat become limited to the extent that utilization of remaining habitat becomes intensive and severe. Turkeys have commercial importance as game and sport birds and may use this habitat peripherally where proper cover occurs on elevated areas.

Winter Residents

Ducks and geese are of major importance because of the drastic build-up of members in fall and winter. The major species are mallards, black ducks, gadwall, green-winged teal, pintail, scaup, and Canadian blue, and snow Geese. Ducks utilize most heavily the shallow coastal waters and marshes and adjacent agricultural lands.

Blue geese winter in few localized regions in the U.S. (California, Texas, and Louisiana in the main). Louisiana coastal marshes constitute one of the prime wintering areas where, together with the snow geese, about 300,000 to 400,000 individuals winter. Loss of large amounts of natural habitat of any of the waterfowl will have considerable impact on adjacent rice and agricultural lands which they will utilize as resting and foraging areas.

Canada geese have resting areas in more northern states. The largest numbers are found in the vicinity of the Chesapeake Bay.

Two species heretofore unmentioned are the bald eagle and osprey. Both of these are secondary and tertiary consumers, and because of their often long or complex food webs, they have been affected over broad areas of North America. Some populations of bald eagles have declined almost to the point of no return. Ospreys also show locally severe reductions. Since both of these species have specific habitat needs and both are of a local, spotty, or sparse nature within this habitat, a rather strict protective control should be maintained.

Migrants

This area provides major resting area for migrants of a variety of kinds, depending on the nature of the cover type. This category is so broad and general that only the most drastic changes in habitat or cover type would have impact, and then the impact would be of questionable nature.

Aquatic Ecology

The value of estuaries and marshlands to the U.S. fisheries is realized when it is considered that almost two thirds (in value) of the commercial catch and most of the sports fish are composed of species who spend at least a part of their life in estuaries. Indeed, freshwater, saltwater, and estuarine fishes taken in amounts of millions of pounds each year.

The thousands of acres of marshes and brackish water areas in the highly productive coast serve as the area of mixing of freshwater runoff with the waters of the Gulf and Oceans. Small tide extremes and intricate water courses combine to produce gradual salinity and water volume transitions, with resultant large zones of relatively stable conditions. These zones may range from quite salty near the ocean to nearly fresh on the inner edge. The wide range of environmental characteristics provides the necessary habitats for many aquatic forms of life.

The most important factor involved in the distribution of estuarine species is salinity. Although organisms typically found in estuaries must have capacities to tolerate changing conditions, many are adapted to function more effectively in given ranges. An excellent illustration of the importance in salinity ranges is that of two enemies of the oyster grown optimally in brackish water areas. Both the oyster drill, *Thias haemastoma*, and a marine fungus, *Dermacystidium marinum,* invade waters of higher salinity during drouth and saltwater intrusions.

General groupings of species comprising the fisheries can be made based on distribution and life histories (with some variation). Four such groupings are convenient and include resident, semicatadromous, seasonal migrant, and anadromous species.

Resident Species

Species dependent on the estuarine environment most of the time and completing their life cycles in this zone are termed resident species. These are represented by oysters, blue crabs, and spotted seatrout, although the latter two species do venture into the shallow waters of the Gulf or oceans. Of the three members of this group, oysters and clams are by far the most valuable of these commercial species which are supported by the estuaries.

One species of fish which does not effectively fall into the resident species or semicatadromous grouping is the bay anchovy (*Anchoa mitchelli*). It is nevertheless an important species to the estuary habitat if for no other reason than its exceedingly large numbers and biomass. Researchers report that this species probably has the greatest biomass of any fish in estuarine waters of the south atlantic and Gulf of Mexico. This species made up almost 85% of the noncommercial vertebrate catch in the Louisiana coastal area. Anchovies are fished commerically off the coast of California. The bay anchovy constitutes a very important food chain organism and must, therefore, be considered when alterations of its environment are contemplated. Its value has been recognized as furnishing food for many predatory species whose habitat distributions suggest its importance in both weedy shallows and surface and bottom waters of open lakes.

Semicatadromous Species

The second and perhaps largest group of organisms are those termed semicatadromous. Species within this group exhibit a generally similar life history which includes

1. Spawning in ocean waters
2. Migration of the young to estuarine areas
3. Growth to subadults in this area
4. Return to ocean waters to complete the cycle

A relatively large number of commercially important species of coastal fisheries are included in this group, and because some return for short periods to the estuaries as adults, they are significant in the inland sports fishery in addition to their commercial value from the ocean.

Important species falling into this group include white shrimp, brown shrimp, menhaden, red drum, flounder, black drum, Atlantic croaker, spot, and sand seatrout, etc.

By far the most abundant species in the commercial fishery in the Gulf area is the menhaden. It is second in value to the two species of shrimp taken. The U.S. is the foremost producer of shrimp in the world, with the shrimp fishery occupying a relatively prominent position in the economy. The shrimp fishery of the Gulf coastal zone consists of two species, the white shrimp, *Penaeus setiferous*, and the brown shrimp, *Penaeus aztecus*. Spawning of both species generally occurs offshore, but at different depths, brown shrimp utilizing waters 240 to 250 ft in depth and white shrimp spawning in shallower waters of 35 to 50 ft. During the months of February, March, and April, postlarvae brown shrimp begin their migration from offshore waters into the estuaries. White shrimp postlarvae are found migrating into the estuaries in greatest numbers during the months of July and August. While in the estuarine nursery area, shrimp feed on a variety of materials, including plants, animals, and inorganic and organic detritus in areas where the substrata is muddy or peaty. Brown shrimp tend to be distributed rather uniformly in waters 2 to 3 ft deep having attached vegetation. White shrimp have a tendency to "school" in waters less than 2 ft deep in areas with large amounts of detritus.

Migration out of the estuary occurs about early spring to midsummer for brown shrimp and around November and December for white shrimp. During their stay in the estuary, growth is quite rapid. White shrimp average about 1½ in. of growth per month, while brown shrimp grow at an average rate of 1½ to 1¾ in./month. Distribution of these species in the estuaries is also governed by salinity ranges, brown shrimp, 9.80 to 60% and white shrimp, 0.45 to 45%. Alteration of coastal areas influencing salinity regimes can be expected to cause shifts in populations of these important species.

In addition to being important in the commercial fishery, most of the semicatadromous species mentioned (with the exception of menhaden) are also valued by recreational fishermen and are taken in large numbers annually. Indeed, it has been reported that the value of the sports fisherman to the U.S. economy system probably exceeds that of the commercial fishery and will in the future. Also, some species of this group constitute the largest industrial fishery in the U.S. for production of oil, pet foods, and fish meal. Although menhaden are the principal component of this fishery, others included in the semicatadromous grouping are sand seatrout, Atlantic croaker, spot, and several other species. It is interesting to note that the menhaden fishery does not conflict with estuarine sports or other commercial fisheries, since the gear used does not select for other species of fish nor does it destroy the estuarine habitat as alleged.

Seasonal Migrant Species

The third grouping of fishes is the seasonal migrant species which appear in the estuarine zones briefly during warmer parts of the year chiefly to forage. Because of this characteristic, they are more valuable recreationally than commercially. Major fishes of this group are tarpon, jacks, king mackerel, Spanish mackerel, ladyfish, bluefish, and cobia.

Anadromous Species

The last major group are termed anadromous. These species spawn and the young spend their early life history in freshwater. When juveniles, they migrate to the ocean where they grow to maturity and return as adults to freshwater to complete their life cycle. Some species, such as Pacific salmon, die after spawning, while others return to freshwater several times during their life.

The major species comprising the anadromous group are the Pacific and Atlantic salmon, steelhead trout, searun cutthroat trout, shad, and several other species of the herring family.

Since these fish spawn in freshwater and spend the greater portion of their adult lives in the sea, the estuary is used mainly as a migratory route between the two environments. However, the time of migration through the estuary varies with the species involved, the temperature of the water, and water flows as well as the geographical area involved. Most species require considerable time in the low-salinity estuarine area to become acclimated to the change in salinity and, in some situations, waiting for appropriate water quality conditions, such as temperature.

ENVIRONMENTAL IMPACTS OF PIPELINING

Pipelines are generally thought to be among the safest means of transporting liquid products. Hazards, however, can be found in stresses imposed by substrate shifts or outside mechanical disturbance. Under water, the latter may occur if the pipeline is snagged by an anchor; on shore, it may occur in excavation activities. Such occurrences can be prevented through the mapping and marking of pipeline routes and adequate burial to prevent interference. If sufficient buoyancy is not achieved, the line can rise and break. This type of casualty can be prevented by good operational procedures and adequate ballasting. In the event of a pipeline rupture, the amount of oil released will be determined by the size of the pipeline, the load it is carrying, and the distance between emergency closing valves. For many pipelines serving offshore terminals, a considerable volume of oil can be released since the distance between valves will be from the terminal to the shore, often a distance of several miles.

It is the intent of this summary to give some general conclusions regarding the environmental, ecological, and cultural effects of pipelining in coastal marshes. An attempt has been made to develop conclusions that are generally applicable to pipelining in the coastal marshes, and not to the effects of a given pipeline. These conclusions will provide the background for the assessment of the impacts of a single pipeline once the route and construction techniques are specified. In addition, several factors bear emphasis so that the conclusions drawn are viewed in their proper context.

The analysis of the effects of pipelining is performed by separating the problem into discrete units. While this is necessary to establish a basis for estimating changes in the environment and the direction and significance of these changes, it is important to remember that the coastal marshlands represent a continuous unit in time and space, i.e., we are dealing with an ecosystem. By concentrating our efforts on discrete factors, we may be ignoring important parts of the system or minimizing subtle but significant interactions.

The high productivity of the coastal ecosystems makes them an asset with regard to all of the areas economic stability and growth. Due to both the ecological significance and the public focus on the coastal zones, it is important that all factors be carefully weighed in further pipeline operations.

The coastal marshlands serve many users — industry, agriculture, fisheries, hunting and trapping, recreation, and navigation. It is difficult, if not impossible, to assign the relative contributions of each of these activities to changes in the marshland environment. Historical patterns or changes are difficult to interpret since they represent the superposition of all of man's activities on the natural variations in the marshlands. The specific environmental effects of pipelining operations can best be assessed in those areas where there is but a single use of the marshes. In those cases, where care have been taken in construction and operation of the pipeline, the environmental effects are minimal.

There is insufficient information available on the environmental effects of pipelining for the quantitative assessment of the impacts of additional pipelining on coastal marshes. No studies have been identified which provide environmental base line data on a single line before construction, during construction, and during operations. This type of data would clearly document any environmental changes in the marshes due to pipelining. Such changes must now be inferred from a variety of data sources.

Physical and Chemical Effects

Changes in the physical and chemical environment from pipelining were identified as the more easily quantifiable of the environmental effects, and, as such, they have received a large amount of attention in the past by study groups and concerned regulatory, management, and conservation groups. It is also evident that changes in many of the biological parameters were directly or indirectly dependent upon environmental changes in the nonliving environment.

Canaling

The soft substrate characteristic of a large part of the coastal marshes results in great difficulty in the employment of standard terrestrial ditching equipment. In all lands which are neither artifically nor naturally drained and with a high water table, either lightweight or floating pipelining equipment must be employed. In either case, a channel is opened which is filled with standing water and will generally remain so unless the channel is refilled.

Land loss due to canaling is a matter of serious concern. Quantitatively, this loss is difficult to pin down. Summarizing, the best estimate from the data available suggests that the direct loss of land from canaling is directly proportional to the width and length of the canal.

For the small shove-type pipeline canal, even when not backfilled, direct land loss is only about 1 acre/mi. For the large 65-ft exploration canals, loss is nearly 8 acres/mi. If a 50-ft flotation barge-dredged pipeline canal is assumed, land loss is about 6 acres/mi (0.01 mi^2/mi of canal). Since coastal area comprises about half land and half water, a pipeline will typically be canaling through marshes only about 50% of the time. Thus, approximately 0.5 mi^2 of marshland may be removed per 100 mi of pipeline.

While not strictly land loss, there is an accompanying loss of availability of marsh resulting from the levee construction with the canal dredging spoils. While the land is still there, it is no longer true marsh and has different characteristics and properties. Reestablishment of vegetation may be a problem in spite of the fact that the marshlands are a high-productivity area, probably because of the reducing nature of these spoils. As noted earlier, the amount of land lost to spoils, banks, and levees is a function of the amount of spoils and their character. As an approximation, the data suggests that the actual width of land whose productivity is altered (including the berms inside the levees) may be five to six times that of the canal.

Thus, for the 50-ft flotation canal example, the diversion of marsh to levee will approximate 12 to 18 acres/mi, and the total marsh area with altered character would be in the range of 30 to 36 acres/mi, in addition to the direct loss of land of 6 acres/mi.

A comparison of advantages and disadvantages of push and flotation canals follows:

Push Canal — Advantages

1. Minimum disturbance to terrain and ecology
2. Minimum land loss unfilled which can be further minimized by backfilling which is often practicable
3. Lower cost where practicable

Push Canal — Disadvantages

1. Slower rate of construction — more delicate control required to push pipe down ditch usually confines work to daylight hours
2. If pipeline has to be lowered for other projects or repairs are necessary, cost savings can disappear
3. Limited access to pipeline for repairs or modification because of marsh condition
4. Water level may drop too low in winter for push construction due to north winds

Flotation Canal — Advantages

1. Adequate clearance above pipeline to accommodate future canals for other pipelines, oil rig access, etc
2. Levees provide resting places and foraging territory for deer and other terrestrial mammals
3. Access to pipeline in the event of repairs or future connections is required
4. In some cases, areas can be opened for wildlife management — possible recreation areas for hunters and fishermen

Flotation Canal — Disadvantages

1. More or less permanent alteration of landscape; greater loss of land
2. Drainage patterns may be changed depending on use of bulkheading and on spoil bank treatment
3. Erosion of land may result if navigation traffic is not prevented
4. Backfilling is impractical since spill that is removed is so fluid it cannot be stockpiled

Canal Erosion

The soft and unconsolidated nature of the soils constituting marshlands leads to erosion when stressed. As an example, many of the barrier islands are deteriorating from natural causes. Erosion of canals is a problem in coastal areas, and the primary cause appears to be boat traffic, though sometimes tidal currents can provide significant scouring action. Rates of from 1 to 2 ft/month have been recorded. This rate of widening is not directly a result of their initial utilization for oil exploration, but rather because they turned into navigation canals offering access to the sea. Canals widen at a rate directly related to canal usage. Several factors influence this rate, of which the most important is marsh characteristics. The softer and more organic the marsh, the more susceptible it is to erosion. Not all of the canal erosion can be attributed to boat traffic. Sometimes cutting a rather small canal will set up a circulation pattern which is self-generating.

Pipeline canals are not generally used as navigation canals; in fact, the prevention of this boat traffic is one of the principal reasons for bulkheading or plugging the canal after completion of a pipeline. (The other is to minimize alteration of existing hydrology and drainage patterns). Thus, pipelining does not appear to be a major contributor to the loss of land due to canal erosion where canals have been blocked.

However, if bulkheads and dams are not maintained, they can wash out around the end, permitting water flow, and inspection and maintenance is required to ensure that these continue to fulfill their designed function. Additionally, where a new canal offers an attractive short-cut across the marshes or to some trapping or fishing area, sabotage is a possibility. Earthen dams have been dug out, wooden bulkheads set afire and burned out, and concrete ones dynamited to open canals to unauthorized boat traffic.

Effects of Spoils

Disposal of the spoil from dredging can have both physical and chemical effects. Physically, the spoil banks or levees change the drainage and use patterns of the marsh. Spoil banks sink by compaction of the marsh upon which they are piled which can alter the existing drainage. Where the spoil banks are continuous, as in a levee, even more interference with drainage patterns is possible.

Even though the elevation of a levee may be only 3 to 6 ft, this is sufficient to increase the drainage of moisture from the soil so that a drier soil results which can result in conditions more favorable for flora other than those of the surrounding marsh.

Turbidity

Estuary turbidities generally fluctuated directly with river discharge, i.e., with suspended solids received and with wind speed.

There are additionally several man-made causes of turbidity: oyster shell dredging, where the turbidity is fairly localized, and shrimp trawling, where turbid tracks extending for miles are visible. While normally conducted in open waters, it would seem plausible that wind and tide could transport these turbid patches well into the estuaries.

Another source noted was that from small boat prop wash, and this appeared in all open water and many of the canals. Maintenance dredging in navigational waters also causes turbidity.

Dredging in connection with pipe laying is a turbidity-causing operation. In severity, this would probably increase from shove or push canal dredging through flotation canal dredging (both of these in the marshes where the spoils are piled on the bank) to channel dredging through estuaries and bays. Clamshell dredges are most generally used in the marshes; in bays the hydraulic or suction dredge is more often encountered. The spoils are pumped out the end of a pipe (commonly 10- to 12-in. diameter), either on to the marsh, on to a "spoil island", along the channel in a ridge, or spread out over a band 200 or 300 ft wide. The advantage of the latter scheme is that existing land drainage current patterns are least disturbed.

Plumes of turbidity trailing away from the operations in the direction of the current. The plume can reach proportions of several yards wide and hundreds of yards long if the substrate is exceptionally muddy. The duration appears to be on the order of several hours at a given location. The concentration and duration of the turbidity caused by pipelining is not expected to produce any permanent damage to the ecology and have minimal impact at time of construction.

Chemical Changes

Changes in pH

Soils high in organic content have low pH values following disturbance. Highly significant negative correlation coefficients are found between pH and organic matter, total carbon, and nitrogen. pH values are lowest in brackish mucks with high levels of sulfates. This undoubtedly ties in with the build-up of reduced forms of sulfur in such mucks; upon subsequent reoxidation, acidity is liberated and the pH is lowered.

Hydrogen Sulfide Toxicity

Hydrogen sulfide is important in marshland ecology. Seawater moving through the marshes provides a continual supply of sulfate ion, the third most abundant ion in seawater. The reducing conditions in marshlands tend to fix this sulfur in three forms: hydrogen sulfide, free sulfur, and iron sulfides. The prevalence of this process is perhaps best shown in the problems encountered in most reclaimed marshlands around the world in which highly acid conditions develop with water table lowering as the reduced forms of sulfur become oxidized. One of these reduced forms, hydrogen sulfide, is highly toxic and levels of concentration could conceivably be high enough at times to kill *Spartina.*

The combination of low pH and sulfide toxicity may explain the slow revegetation of levees formed from the subterranean muck excavated in dredging a canal since oxidation and neutralization appear to proceed slowly. The artificial hastening of this neutralization may offer possibilities of reducing the time for revegetation.

Dissolved Oxygen

Dredge spoils are high in biological oxygen demand (BOD) and chemical oxygen demand (COD), and the presence of sulfides is frequent. As an example of the response of a closed system, the dredging of a pond for some mariculture experiments at Nicholls State College, Phidobaux, La. is illustrative. The water in a 30-ft-diameter × 10-ft-deep pond dug in a peat marsh had a reading of zero dissolved oxygen in May when completed, and it was not until September (4 months) that the oxygen content stabilized at normal levels. These results cannot be extrapolated directly to pipeline canals due to the differing circumstances. The marsh was peaty, there was no exchange with any other water body, and the surface to volume ratio was low. No data on dissolved oxygen content associated with pipeline canaling have been accumulated, but they would not expect to be as low as in this special case since the canals do have exchange paths with other bodies of water and the surface to volume ratios are generally higher. Any low oxygen content would be expected to be only a transient condition.

Salinity

Salinity, important to the estuarine nursery area, can be affected by numerous activities of man, including dredging for oil well exploration, pipeline laying, navigation, reclaiming land, and flood control.

The importance of salinity results from the varying tolerances of different members of the ecosystem. Salinity tolerances of some significant species are

1. Oysters cannot grow well below 5‰ salinity, and oyster predation becomes of major significance in salinities of about 12‰.
2. Shrimp prefer salinities of 8‰ and above.
3. Optimum requirements for fur animals and waterfowl are probably below 5‰.

One of the most obvious and fundamental characteristics of any estuary is the gradation and orderly fluctuation of salinity in the zone of saltwater-freshwater mixing. In the open bays at the seaward ends of estuaries, salinities change rapidly in response to winds and tides, whereas in the brackish zone, salinity fluctuations tend to be muted by the complexity of tidal channels and passes connecting these areas with the open bays and sea. The shallow nature of the estuaries promotes rapid vertical mixing so that there is generally little change in salinity from water surface to lake or bay bottom at a given time. It is recognized that in deeper channels a strong vertical component (gradient) may be present. Deeper channels serve as avenues of salinity intrusion into the estuaries.

The relationship between pipelines and salinity changes is difficult to quantify. Where a flotation canal, 40 or more ft wide, traverses marshlands and embayments without regard for changes in natural drainage patterns or for the disruption of currents in bays and water flows in marshlands, changes in salinity can be expected to result. However, this effect upon the salinity regimes would seem to be more associated with canals which are not bulkheaded or dammed, i.e., access rig canals and channels, especially, navigation canals.

Additional Impacts on Water Quality

In addition to the impacts involving waters in estuaries and marshes, general areas of environmental concern for water quality associated with dredging activities are those related to basic changes in the water circulation pattern and its implications, interference with groundwater quality and/or quantity, the effects of associated turbidity, and the effect on the speed, volume, and direction of both surface and bottom currents.

In most cases, water circulation pattern change is likely to be the most significant impact of the dredging activity, both in time of duration and ramification. The deepening of near-shore areas offers a potential threat to freshwater aquifers. The decrease in substrate which separates aquifers from overlying saline water could allow mixing of the two either within the aquifer or, if freshwater escaped, within the region alone. In the former instance, the impact would be felt on the quality and quantity of groundwater; in the latter case, the impact would be on the salinity regime of deepwater areas and, consequently, on the indigenous biota.

Nutrient or chemical-rich dredging spoil can have significant effects on a disposal area. These can take the form of an increase in nutrients and a concurrent spurt in area productivity or a destruction of local organisms through the introduction of toxic materials. Such impacts are likely to be extensive in both time and area as these materials are slowly dissolved and spread through the water column. Onshore disposal of spoil can result in a leaching of chemicals into groundwater areas as well as adjacent surface waters, thus having implications for human consumption of these waters.

Structures built on pilings, such as booster stations, through which water may flow, serve to alter the movement of swells and waves, especially if the above-water solid portion of the structure is low enough to impede wave movement. If this is the case and wave movement to the shore is altered, the effects may be significant. If the shoreline is used for recreation, the lows of surf may reduce its desirability. The loss of wave action could also significantly alter the movement of sand to and from the beach.

Long structures with an extensive string of pilings may also alter the water circulation. While little is really known about how circulation and flow will be affected, there appears to be evidence that the bridge tunnel across the Chesapeake Bay has substantially reduced water movement in the upper part of the bay. So this possible effect should not be overlooked.

A solid offshore structure such as a filled island, breakwater, or jetty will definitely have an impact on the movement of water. Just what the impact will be depends on the location of the structure, its size, and the character of the water movement before construction. A solid structure which extends above the surface will stop passing waves and prevailing current and water circulation will be altered. Just how the circulation pattern will be changed is dependent upon many variables, including the tidal flow and tidal prism, water depth, size of the water body, and water inflow.

Secondary effects of changes in circulation and wave action may also be important. Changes in water flow in partially enclosed bays and estuaries may also alter the salinity and water quality since a decrease in flushing action will tend to lower the salinity of a bay's waters while decreasing the rate at which pollutants are removed to the ocean. Such changes in water movement may also significantly alter the way that sand and other solid materials are transported and deposited. Resultant changes can involve loss of beach where the sand will normally be deposited and deposition where there ordinarily will be little or none. This can have important implications for recreation and shoreline residential uses. Additionally, it can cause increased sedimentation in channels, resulting in an increase in maintenance dredging or loss of channel.

Biological Effects

The terrestrial environment for the purposes of this report includes all lands and lands supporting terrestrial and emergent vegetation. Although the boundaries between lands and the estuarine system are not always clear and the importance of the terrestrial energy system to the estuarine is immense, the separation permits categorization of ecological conditions and changes associated with pipelining. On a subject basis, consideration is given to plants, animals, birds, insects, and rare and endangered species.

Construction on Plant Communities

Plant communities of the coastal marshes and swamps are highly productive and support a high diversity of consumer types both terrestrial and aquatic. The primary direct effects of pipeline activities on the communities is somewhat similar for all types studied. These effects include loss of vegetation in the line of the ditching or dredging activity and subsequent loss of vegetation through erosion of canal berms, compaction of vegetation and soils along areas traversed by surveying and construction equipment, and burial under spoil banks.

The extent of vegetational loss is directly related to the width of the ditch or canal, i.e., 0.01 mi^2/mi for a 50-ft wide canal, and, of course, the larger the canal, the greater the loss. The larger open canals in areas of very unstable soils subject to commercial and recreational navigation experience significant erosion from prop wash and wave action.

Loss of vegetation by spoil and levee coverage is likewise directly related to the aerial extent of spoil spreading and is probably not greatly dependent upon method of placement. An exception may include pump dredging, however, little documentation is available on the effects of the technique.

In cases where very low marsh lands contain freshwater or brackish water and are protected from tides and flooding by natural or man-made levees or other features as vegetation, etc., canaling may contribute to regular or periodic flooding by more saline waters than previously experienced. Such environmental changes result in vegetational changes ranging from small shifts in species to complete loss of living vegetation through "die-back".

Conversely, canals which capture freshwater drainage may result in transport of these waters into and on lands usually subject to high salinity concentration. Changes in plant species diversity (species composition) will occur in all areas where there is either a major or minor change in the environment. Major changes as land excavation, drainage, or flooding result in a complete change in the kinds of organism occupying an impacted area. Such changes are usually undesirable unless the purpose of the environmental alteration is for management for maintenance of a particular likewise productive ecological system. Minor effects result in slight shifts in productivity and concomitant shifts in species occupying an area. Thus, slight changes may be either desirable or undesirable, depending on the type of effect and vegetational shift. For example, changes which reduce productivity, species diversity, or availability of detrital material to aquatic environment are detrimental. Beneficial changes include increase in availability of detrital material, increase in habitat and species diversity, and increase in productivity. It has been noted that the stabilized spoil banks supporting shrub and tree vegetation provide food and shelter for both deer and rabbits, but do not necessarily support the original vegetation and associated animals. This effect is particularly noticeable on protected lands.

The marsh and swamp lands are quite rich in dominant and successional plant species. This richness contributes to the dramatic resiliency of the systems to disturbance from natural forces and man-made activity. Because of the dynamic nature of the systems, no effects of pipelining were identified for rare and endangered plant species.

Additionally, the lowering of water tables via ditch and canal drainage of lands slightly higher than sea level may allow better aeration in the soils as well as spoil banks, resulting in reduced forms of sulfur being oxidized. The resultant sulfides and acids, in some marshes, overwhelm the buffer potential of the salts and may cause extensive vegetation damage. The extent of the damage to vegetation is related to the extent of the washoff from spoil banks and the distance from the pipeline ditch that the water level has been lowered. As yet it is not possible to adequately predict the location and possible extent of such occurrences from the information reviewed. However, where sulfide toxicity to vegetation does occur, it is readily noticeable. Inputs of these toxic materials as well as contaminated dead vegetation become important with respect to effects on the aquatic environment.

The significance of the category of plant communities is perhaps the most complex, but the more easily explained as compared to other ecological components of the marsh ecosystems. The marshes vary in productivity, but may produce up to 15 ton/acre/year or 5 to 10 times the average productivity for crops in the agricultural U.S. Approximately 10% of this productivity is consumed by bird, mammal, and insect grazers, leaving 90% or over 13 ton/acre as an energy input as dead plant materials to be accumulated as organic matter on and in the soil, consumed by bacterial and fungal decomposers, or as a detritus food source for estuarine organisms. The proportional quantity available for the latter is highly variable, but of no small amount. It is these materials that contribute to the protection of food and food chain sources used by 95% of the estuarine fish and shell fish. The marshes are also the habitat for the myriad of organisms which either feed, nest, or rest in the marshlands.

In context of the impact of pipeline disturbance to marshlands, a unit area of vegetation loss is equivalent to five to ten times the equivalent amount of productivity that would be disturbed within the agricultural lands of the central U.S. The nature of the vegetation patterns are largely controlled by water depth and salinity. Slight changes in either of these results in a change of the vegetational type either to more advanced stages, i.e., shrub, or to mud flats if an early stage is subjected to extremely undesirable stresses. As little as a 2-in. change in elevation will result in a shift in vegetation.

In order to quantify and put in better perspective the effects of pipeline activities on plant communities, more detailed data are needed regarding magnitude and importance of effects of pipeline ditches and canals; scientific observation needs to be undertaken in the marsh types occurring in the coastal area. Specific data are needed relative to

1. The direct losses of vegetation in terms of species and productivity from pipeline placement
2. The indirect effects from water level changes
3. Vegetational effects from salinity changes
4. The nature and progress of revegetation of spoil banks.

The technique of ditching without backfilling whereby land is converted to a definite water form is relatively final with respect to the loss of vegetation. However, in certain cases where the ditch is small and berms unstable, the ditch may in time become revegetated.

The minimization of effects on the plant communities can only be accomplished by detailed studies along potentially proposed pipeline route alternatives. Therefore, careful consideration need be given to the characteristics of the physical and chemical environment, the vegetation type, along with the possible types of ditching that is feasible to employ. Consideration should also be given to the expense and time effort that may be required to reclaim (revegetate) spoil banks.

Oil Spills on Plant Communities

Studies on the effects of oil on marshes indicate that oil adheres to plants and little washes off during successive tides. Under the oil films, leaves may remain green initially, but eventually they turn yellow and die. Plants, however, recover by producing new shoots, a few of which are observed within 3 weeks of pollution, unless large quantities of oil have soaked into the plant base and soil. Seedlings and annuals rarely recover directly. Chronic pollution may completely eliminate vegetation.

The effects of oil on plants depend on several factors, including

1. Species and age of the plants
2. Time of year and whether the plant is in an active growth stage or dormant
3. Amount and type of oil involved
4. Degree of weathering of the oil

Weathered crude appears to be less toxic to salt marsh vegetation than fresh oil. A badly oiled marsh at Bentless on Pembroke River which was severely damaged after the Chrysse P. Goulandis tanker accident in 1967 was virtually completely recovered 2 years later. Another study of marshes, in Brittany, damaged by oil in April 1967 showed that these marshes were able to withstand heavy oil concentrations with only slight and probably short-term floral composition changes. The oil had been at sea for 14 to 18 days and essentially all toxic fractions had been lost. Within 16 months there was extremely vigorous growth of certain plant species, and it seemed that these plants may have derived some actual nutritional benefits from the breakdown products of oil.

In general, it appears that heavy concentrations of oil on the marsh areas will temporarily destroy the vegetation. However, based upon the limited data available, the oil does not appear to cause any significant long-term damage to the marsh area.

Chronic, low-level oil spills may impose greater environmental damage in estuaries and marshes because of mixing with pollution from other sources. Thus, additional stresses caused by chronic oil spills will likely degrade water quality and have long-term toxic and subtoxic effects on the marine environment.

Construction on Animal Populations

Populations of native and introduced game and nongame animals respond in numerous complex ways to environmental alteration. They are dependent upon the characteristics of and change in the physical and chemical environment and as consumers are dependent on the vegetation which supports them.

Considering these factors and the adaptability of the animals themselves to environmental change, the effects of primary importance are direct loss of habitat lands whether temporary or long-term, dissection of terrestrial populations, and behavioral disturbance during construction.

Of particular interest is the loss of habitat land for animals of economic interest as the furbearers. This loss consists of both that disturbed by the ditching operation and the uncolonized spoil banks. On the other hand, considerable benefit has been realized historically through the use of open canals as access to lands by trappers and hunters. The magnitude of the effects and benefits has not been specifically evaluated such that values of importance can be assigned.

A similar effect of reduction in habitat area has been identified for other terrestrial animals including game and nongame animals.

It is expected that ditching may, in some cases, act as a migrational and home range barrier for some terrestrial species. However, most of the organisms occupying the wet marshes are well adapted to crossing open barren water areas. The crossing of spoil banks would, however, subject many animals to increased predatory vulnerability. Very little may be generalized concerning these and other behavioral responses in nature except for selected species in better studies areas.

For animals which spend a large part of their existence in the aquatic environment, principally reptiles and amphibians, there is an increase in habitat which is directly proportional to the amount of water surface area gained. Species using new water areas are quite numerous and probably variable from area to area. Specific line consideration would include size of canal salinity, surrounding vegetation type, etc.

Pipelines crossing habitats containing protected species (such as alligator) require special attention relative to loss of habitat or dissection of populations. Increasing water surface area would increase habitat for the alligator, but there may also be an increase in availability of habitat areas to poachers. The effect of poaching is outside the responsibility of the pipeline industry per se.

Another important area of concern is behavioral influence during pipeline construction and surveillance. In most cases, displacement and interruption of activities of animals would be temporary, with little or no continued impact in backfilled canals. However, multipurpose canals would result in animals being periodically subjected to noise and sight of moving objects which may affect the local population success for some species. Additionally, aircraft may elicit a significant amount of avoidance behavior in both mammals and birds when pilots bring their craft near the ground. A certain amount of this effect is unavoidable, however, in the construction and operation of pipeline services.

Similarly with plants, animals experience an increase in habitat diversity where stable spoils become revegetated with shrubs. Although the new vegetation may not be as the original, it provides shelter and possibly food, particularly for deer.

The significance of this category is difficult to assess except for the importance for harvest of fur-bearing and game animals for which data are presented in the introductory descriptive sections. The change in environment resulting in changes in species and densities of animals is perhaps less important than changes in vegetation because they consume a lesser proportion of the fixed energy in the ecosystem than that consumed by the decomposers and perhaps the estuarine organisms. The parameters of species composition and density are, however, good and, in many cases, more sensitive indicators of ecosystem stability than many other of the categories.

Further documentation is needed concerning these population shifts, particularly with respect to vegetational changes associated with varying pipelining modes in varying marsh types.

Oil Spills on Animal Populations

There is a distinction between the physical effects of a coat of oil on an organism and the effects of the hydrocarbon components of the oil on the physiology of the organism. Toxicity refers to the effects of one or more hydrocarbon fractions which disrupt the cellular or subcellular functions of the organism and cause death if the disruptions are widespread and severe enough. This implies that hydrocarbons come into contact with the cell membrane and either pass within the cell to disrupt metabolic processes or disrupt the function of the membrane itself. Coating affects ecosystems or organisms. For example, a coat of weathered oil covering the respiratory apparatus of an organism will not harm individual cells by contact, but will kill the organism by suffocation. Death in this case is not a function of the coating agent being a hydrocarbon mixture (weathered oil has lost most of the fractions which are harmful at the cellular level); the same effect could probably be induced by using the same volume of glue.

In terms of respiration and feeding interference when considering littoral or benthic organisms, a division can be made by differentiating between sessile and mobile organisms. Whether or not an organism can move away from oil deposited on the bottom plays a large role in determining whether it survives or not. Even if an organism cannot move and becomes covered with oil, it still may be able to protect itself (e.g., bivalves) until the oil is removed or until the oil layer becomes shallow enough for the organism to break through to clean water.

Apparently marine mammals fail to recognize an oil slick until coating is inevitable. These mammals must periodically surface in order to receive air and in doing so they are in danger of being coated with oil. One dead dolphin found washed ashore in Santa Barbara reportedly had its blowhole plugged with oil, although this finding has been disputed. Also in the Santa Barbara spill, many weaned elephant seal pups were coated with oil and apparently were not harmed, although immediate mortality could have been much higher had the oil spill taken place at the time when the pups were feeding. There were reports of a tripling in the number of premature births of sea lions which, like the sea elephants, have rookeries on the islands in the Santa Barbara Channel.

It seems probable that marine mammal mortalities can only be indirectly related to coating. Warm-blooded marine organisms must maintain their body temperatures at a higher level than the surrounding water. This requires an efficient insulating layer surrounding the body. Birds use a layer of waterproofed feathers, while mammals usually combine layers of fat with a thick coat of fur as insulation. A coat of oil can significantly change the insulating properties of the fur, perhaps causing the animal to lose heat and lower its resistance to disease. This is just one of several possible hypotheses that can be used to indirectly connect oil coating with marine mammal mortality, although, based on current evidence, none of these hypotheses has been proven conclusively.

Due primarily to a high degree of mobility, most forms of land animals may escape the effects of oil spills and onshore leakages. These animals, including most mammals and reptiles, can escape all but sudden spills on land and can avoid water-borne oil easily. On the other hand, the mammals, reptiles, and amphibians living in aquatic, swampy, or marshy habitats are not so fortunate as their natural habitat can be destroyed by the oil so that they will be unable to return. Those animals which do not flee from the oil may be killed by the ingestion of contaminated foods or suffer "skin burns" from being covered with oil. If they flee from their homes, these animals face stress and possible death from overcrowding, starvation, predation, or disease.

The terrestrial mammals which would most likely be affected by an oil spill and which are of some aesthetic or economic value to man are the river otter, beaver, and muskrat, etc.

Construction on Birds

As has been pointed out, the prime ecological consequence of pipeline construction and operation results from such physical alterations of the habitat from channelization and associated changes. In general, drainage patterns and water levels are changed, salinity is altered, actual marsh habitat is lost or altered, and beach or sea rims are altered by lowered or raised water levels. These topics will be alluded to in terms of general or specific effects upon groups of birds or individual species.

Herons, egrets, and owls are colonial species. Thus, the colonies are local or spotty in nature. Bird behavior involves a great deal of tradition in the historical sense. Once traditional or established colonies are lost, say to erosion, it is uncertain how satisfactorily the species adjust to other regions. The colonies are where they are, in part, because of locally abundant food supplies. Changing drainage patterns, salinity content, etc. may alter the distribution of the food supply in such a way that even if the physical habitat for the colony remains they become unusable without food. There is no fear of herons and their allies invading agricultural lands where they might cause economic loss should their habitat be reduced. Rather, there will simply be a loss of numbers of individuals. One of the obvious impacts is the effect of a changing saline content on the birds invertebrate food base. Often the food web is complicated and dependent on such limiting factors as pH of the water or turbidity.

The physical conditions of water depth is often critical to fish, other vertebrates, or invertebrates used by birds. Fish may become unavailable as a food source if water becomes too deep. Further, much of bird behavior is dependent on energetic considerations. If changing habitat removes the food source too distant from the nesting sites and it becomes too costly calorically to exploit the food source, then nesting sites are often abandoned. Pelicans, cormorants, herons, etc. are species which show such responses.

Rails form a huntable economic resource. Density and nature of the vegetation appear to limit or determine the distribution of rails. The nature of the vegetation is in turn dictated, in part by the pH or depth of the water. Thus, it is expected that rails will respond in a negative manner where marsh is converted to water.

Pelicans show the greatest productive fitness when in seclusion. The channeling opens the habitat to boat traffic, and the physical proximity of boats or noise produced by them is known to keep pelicans away from their nests. The result is the exposure of eggs to environmental adversities (such as heat or predation) and, consequently, decreased nesting success.

A major impact with economic implications is the changing of feeding habits for waterfowl. With the elimination or alteration of the *Spartina* or *Scirpus* marshes, waterfowl can be expected to utilize more heavily agricultural lands.

Another complicated association exists with the endangered peregrine falcon and its food supply. This endangered species seemingly requires an abundance of the proper size range of food species durings its migrations. These food species congregate along coastal marshes and sea rims. In many coastal areas in other regions of the migratory routes, homes now occupy the essential environment, and habitat diversity may be reduced. It is essential that the proper habitat for the prey species be maintained in sufficient quantity to help eliminate competition and starvation where these birds are forced to congregate locally.

For the birds in general, then, the responses to habitat perturbation will vary as covered above from essentially no measurable response, such as in many migrant and resident passerines and slight shifts to agricultural lands for feeding and nesting, to total abandonment or exclusion of the habitat resulting in mortality and loss of individuals.

The value of the coastal marshes to many diverse groups of resident and migratory birds is hard to express in any quantitative manner, but it is evident that the carrying capacity of marshes is important to the continuance of the avifauna population.

Birds perhaps more than any other kind of organisms are able to rapidly disperse when disturbed. Except for studies of nesting success, very little is conclusively known as a general basis of ultimate effect of slight changes in landscape utilization and short-term dispersal of birds. More direct observations are required from pipeline operations. These studies could be incorporated into routine studies on the effects of existing pipelines on the environment and the effects of surveillance on numbers and kinds of birds dispersed by boat and aircraft passes.

Effects of Oil Spills on Birds

There is almost unanimous agreement that birds have probably suffered more from oil spills than any other groups of organisms. Bird casualties from oil pollution number many thousand each year.

Birds are warm blooded and thus must insulate themselves against their surroundings. Marine birds have an additional problem because their insulation must also be waterproof. Aerial species rarely come into contact with oil, whether it is floating or washed onto the beach; thus, they are not especially endangered. Swimming species are continually subject to contact with floating oil slicks. Often their first reaction upon coming in contact with the oil is to dive beneath it, invariably resurfacing in the oil, and thereby, compounding the problem.

Those orders of birds which would be most affected by oil spills are the loons, grebes, tubenoses, pelicans, herons, waterfowl (ducks and geese), and cranes.

Two factors that appear to be involved in the death of aquatic species — the external oiling and the ingestion of the oil — have received some attention during the past few years. While the chemical differences between commercial crude oils and between various refined products present problems in determining the exact effects of the petroleum on the birds, some reports reveal that lipid pneumonia, GI irritation, fatty livers, adrenal cortex hyperplasm, acinar atrophy of the pancreas, and toxic nephrosis have been found in several kinds of ducks when they were fed industrial oils, cutting oil, or diesel oil. This suggests that the preening of oil-fouled feathers, which results in the ingestion of large amounts of oil, causes the death of many birds.

Furthermore, the external oiling of aquatic birds has numerous direct negative results. The oil coats the bird's body, mats its feathers, displaces air between the feathers and the skin, and breaks down the natural insulating oils and waxes. Damage to the plumage, the weight of the fouling oil and water, and the loss of natural buoyancy prevents the bird from flying and necessitates constant swimming. Exhausted and emaciated by constant swimming, an increased metabolic rate, and improper feeding, the oiled bird also spends a disproportionate amount of time preening (instead of feeding), and its condition worsens. Efforts to preen the oil from the feathers results in the ingestion of the oil which is usually toxic to the waterfowl. Thus, the waterfowl, weakened by the combination of a number of factors, usually succumbs to the adverse effects resulting from its contact with oil.

Efforts to rescue oil-polluted birds have achieved negligible results, with rarely more than 10% of the treated birds surviving. For example, of the 7849 birds rescued after the Torrey Canyon spill, less than 400 (0.05%) survived. Those birds surviving the initial treatment for ingested and external oil pose additional problems of housing, feeding, and handling. Also, crowding, behavioral stresses, and diseases can create numerous difficulties. Because of these factors it is estimated that it can cost $1000 to *try* to save a single bird. With a success ratio of only 10%, only 100 of every 1000 oiled birds treated are likely to survive; as a result, the cost could run to $10,000 per surviving bird.

Because the costs of saving waterfowl are so high, it is necessary to find ways to keep the birds out of the oil when a spill occurs. This is complicated by the fact that some researchers think birds are unable to distinguish oil-polluted waters from unpolluted waters. Others think seabirds are attracted by dead and dying marine life on oil slicks and oil-covered beaches. It has been suggested that the bird's own alarm call or other sounding devices be used to scare birds away from oil slicks. However, these devices, which have been used to scare gulls and rooks, may not be applicable to all species of aquatic birds.

Invertebrates

Mites, grasshoppers, snails, etc. are the primary consumers of the terrestrial vegetation and decaying vegetation products in coastal marshes. Although little information was reviewed concerning effects of habitat alteration on these organisms, observations were found for four marsh plant association types. They were *S. alterniflora, D. spicata, S. patens,* and *Scirpus robutus.* Average reduction in invertebrates by plant associations were 48, 80, 80, and 91, respectively. Invertebrate groups included mollusks, crustaceans, spiders and mites, leafhoppers, beetles, and miscellaneous. Losses as high as 94 and 98% were reported for mollusks and crustaceans during sampling periods for all marsh types. Reduction in numbers of species ranged from 15 to 25 in the study areas. The losses were attributed to drainage on loss of habitat and increased predation.

The significance of this category is twofold. First, these organisms are responsible for breaking down plant materials so that they are utilizable by decomposers and estuarine organisms. Secondly, the organisms themselves serve as important food chain components to estuarine organisms and to waterfowl.

Field data are needed on losses of these organisms along representative marsh pipelines for both spoil banks and surrounding marshes where groundwater or standing water has been altered.

Construction on Aquatic and Estuarine Biota

Earlier mention was made of some of the general effects dredging or similar activity may have on environments and biological populations in coastal zone. However, it is necessary now to examine some of these effects as they relate to specific life forms of both ecological and (ultimately) economic importance. Although most of the changes which can occur have effects on the total ecosystem by altering their environment either directly or indirectly, most of the following discussions will center around indirect effects on the more valuable species of the commercial fisheries and habitat.

Salinity Effects

The coastal zone is one of the most productive fisheries areas in the world and indicates a correspondingly high primary production. Studies of community nutrition in representative areas of the coastal zone have indicated organic detritus as the most important contributor to estuarine production. Species reported as chief contributors include the two grasses, *S. alterniflora* and *S. patens.* Seasonal death and decomposition of these producers result in a springtime release of nutrients and detritus at a time when various stages of estuarine-dependent species are returning. Organic detritus thus made available to consumer species initiate the eventual fisheries production for the ensuing year. For example, species of fish and invertebrates in which organic detritus can constitute 50% or more of the food include large-scale menhaden, striped mullet, marsh clam, Atlantic croaker, white shrimp, sea catfish, river shrimp, gizzard shad, and hogchoker. Through the support of a large number of organisms in the estuary at this lower trophic level, marsh vegetation serves as a most important step in overall fisheries production.

The interference of drainage caused by deposition of spoil banks may retard normal flushing, and hence, normal distribution of valuable nutrients in the area, and may cause waterlogging of sometimes large areas of marsh. The latter has been regarded as a probable cause for *Spartina* dieback due to increase in salinity in the poorly drained areas and, consequently, a loss of food energy to estuaries. *Spartina* could be affected by poor drainage producing

1. Chlorosis from lack of available iron
2. Hydrogen sulfide toxicity
3. Oxygen deficiency in the roots

Where dredging activity produces drainage of marsh areas, loss of nutrients to the community may go beyond the primary level. The effects of ditching in tidewater marshes was found to significantly decrease invertebrate populations in both *S. alterniflora* and *S. patens* associations after ditching. Estuarine organisms, notably mollusks and crustaceans, were most drastically reduced in number in the ditched zones of both associations. Losses of these two groups after ditching in the *S. patens* may be as high as 95%, with almost 100% in the *S. alterniflora* zone.

Perhaps the one environmental parameter which has the greatest influence on the aquatic ecology of the coastal area is salinity. Alternations of drainage patterns which may cause this parameter to fluctuate or totally change will have an effect on the production of a number of organisms. Two of particular importance are the brown and white shrimp. Salinities at which one can find these organisms in the greatest abundance is more limited than the range in which they can live. Also, as the postlarvae of these species migrate from offshore, the absence of optimal salinity (and temperature) within the estuaries can result in a slow growth and increased mortality.

Of the various types of man-induced modifications of the estuarine environment as a habitat for the young of economically important shrimp, many resulted in salinity alterations to some degree. However, salinity changes appeared to have an indirect effect on the environment for young shrimp more from a loss carrying capacity of the marsh areas through reduction of plant cover and food sources.

Direct effects of salinity on oysters are not as pronounced as indirect effects because of their wide tolerance range of from 7 to 30‰. However, too much freshwater into an area can cause sudden and catastrophic results. As has already been mentioned, salinity changes reaching oyster-growing areas of an estuary can create conditions favorable for two natural enemies of the oyster: the marine fungus, *D. marinum*, and the oyster drill, *T. haemastoma*.

The blue crab population can be affected by changes in salinity both directly and indirectly. Larvae of this organism spend some time in the low-saline marsh areas, and thus, changes in salinity can affect the distribution. An indirect effect can be found in alteration of the available food for the adults of the species.

One important finfish of the commercial fishery, the menhaden, has been reported to be salinity sensitive during its early life. Young larvae develop normally only in very low salinities until past a certain stage of growth. If they are exposed to high salinities before this stage, they develop abnormally.

Turbidity and Siltation Effects

A second factor which is a direct product of dredging activities is the production of turbidity and siltation together with a possible alteration of water quality.

Studies on the effects of dredging on finfish show no damage to fish even in the close proximity of the dredge. There is some evidence to show that organic detritus is stirred up, producing a beneficial effect on both fish and shellfish. The natural reaction is for the fish to avoid the turbidities if possible. If fish are removed from turbid waters, they are able to dislodge and remove clogging mud from their gills. Fish have been killed in the laboratory, but those fish that died from high concentrations of suspended solids were found to have experienced mechanical interference with normal respiratory exchange through the gills.

There appears to be no gross effects from disposal of fine materials on fish eggs and larvae or on adult fishes which are caught or held in cages near the area of operation. No effect on the phytoplankton and zooplankton has been demonstrated, but significant loss of bottom organisms in a wide area around the operation can happen as a result of smothering. However, some species were able to survive and repopulate some areas soon after materials deposition. The main effects of suspended matter on fish are an interruption of the food chain as a result of reduced photosynthesis by food chain and food chain-consuming organisms.

Effects of turbidity on shrimp as a direct result of dredging are not known. More indirectly, the resuspension of nutrients could be beneficial, whereas resuspension of toxic materials in the sediments would be negative.

The greatest concern over silting effects has been in regard to the oyster industry. This has been because of the sessile nature of this valuable species and its inability to avoid unfavorable conditions and because by far the highest production of oysters is on private leases. Alleged damage to these private oyster beds as a result of dredging activity in the area is commonplace, and payment for damages is generally in excess of actual value. This is due to the claim of many growers that a larger number of seed oysters are planted than actually are.

Needless to say, a number of studies have been made in regard to silting effects on oysters. Results of these studies showed that oysters are able to survive dredging operations even when exposed to the mud and silt of the cutter. Only when oysters were covered or mostly covered with spoil discharge do they fail to survive. Although turbidity and siltation may not cause death of oysters, they can have an effect. When oysters were placed in turbid environments, the pumping rate was reduced by as much as 50% of normal. This effect was noted in the presence of 0.1 g/ℓ of suspended solids.

Observations and Problems

A number of direct and indirect effects which are related to dredging activities in this area have been presented. However, in the final analysis, modification of the environment which causes physical loss of habitat (either directly or indirectly) and change in salinity regimes stand out as the most important and dredging affects both of these.

If, indeed, modifications of the environment must be made, every precaution should be taken to minimize the impact. Each area in the coastal zone has its own characteristics and must be approached in regard to modifications on an individual basis. Appropriate timing of alternative operations in this environment and proper consultation during operations is one way to minimize effects. Based on information available now, any such activity would have its least effect on the biology and ecology of the area when carried out during the winter months. However, there is no way to escape the fact that where there are alterations, biological changes will occur.

As is readily apparent from the review of aquatic and estuarine responses to the pipelining activities, there is a distinct need for specificity in information. Much of the present information relates to total changes in the wetlands without regard to specific causative actions which promoted the responses. Many of the problems can be readily solved by field programs designed to obtain data relevant to actions by the pipeline industry. Particular emphasis is suggested for the areas of:

1. Toxicity in sediments
2. Magnitude of impacts resulting from various pipe-laying modes
3. Effectiveness of backfilling and bulkheading in minimizing aquatic and estuarine effects

An equally important consideration in establishing perspective for the pipeline industries is that of existing effects resulting from other marshland users and possible interactions that new pipelining efforts would have with the other users. Many of these problems can be addressed more specifically when specific routes are selected in the future. The nature of possible interactions are difficult to generalize at this time because of the geographic variation in the marshes themselves and the nature of distribution of other causative changes.

Oil Spills on Aquatic and Estuarine Biota

Knowledge of the biological effects of petroleum spilled in the marine environment is still incomplete. Most of the available information comes from studies of the several large oil spills from tanker casualties such as the Torrey Canyon. There are relatively few well-studied cases, and the results from them are not in complete agreement. Still it is possible to identify some of the major impacts of a spill that act individually or in combination and to examine these effects on the various marine organisms.

Regardless of the source and size of the oil spill, several biological responses are important:

1. Immediate lethal toxicity
2. Lethal or sublethal effects of direct coating of marine life by oil substances
3. Altered behavioral activities
4. Incorporation of aromatic hydrocarbons into the food chain
5. Changes in habitat, especially for attached organisms, due to the deposition of oil on rocks, sediments, or other habitat areas

In general, because of high dilution factors and widespread biotic populations, offshore areas have greater assimilative capacities and are relatively less vulnerable than shoreline or estuarine areas to critical environmental damage from chronic low-level spills.

The most important effect of oil on marine organisms, from the public health point of view, is incorporation of hydrocarbons in food chains. There is ample evidence that a variety of marine organisms do, in fact, incorporate and accumulate hydrocarbons. Incorporation of hydrocarbons themselves apparently does not affect the organisms directly. However, two potential problems are important to man: (1) accumulation in organisms of polycyclic aromatic hydrocarbons, especially carcinogens, and (2) tainting of edible species.

Finfish

Finfish and other mobile pelagic organisms which can detect low concentrations of oil in seawater presumably avoid higher concentrations. The eggs, larvae, and juvenile stages of these fish are generally more susceptible to the effects of oil than are the adults.

Whereas fish can avoid oil films on the surface, damage may result from oil products emulsified in seawater. Damage is possible from physically clogging the gills, precipitation of certain metals on the gill epithelium, and direct action of hydrocarbons on sensitive gill cells. Emulsified oil may also block taste receptors or mask the natural chemical messengers important in the biology of fish life.

It has been suggested that components of oil might disrupt the behavior of all marine organisms, including their mating and reproduction, by interfering with their behavior patterns. Although this seems quite possible, there exists no direct evidence for it. It is considered likely that gradual accumulations of oil could definitely prove to be a limiting factor in terms of spawning, migrating, and foraging habits. Thus, a gradual reduction of finfish activity could be associated with regular low-level oil spills occurring in the port vicinity.

Some concern has been raised that oil spills result in cancer to humans or valuable marine life, but no direct evidence has been found to confirm this. Known carcinogens exist in crude oil, fuel oils, and waste oils, and these have been detected in marine sediments and in the tissues of barnacles and oysters.

Moreover, fish with a high frequency of lesions, possibly cancerous, have been observed in areas exposed to oil pollution. Direct proof that these lesions were caused by oil components or that carcinogens derived from spilled oil reach man or other valuable marine life in hazardous concentrations is lacking, but the possibility seems apparent.

Benthic Fauna

The benthic fauna may be divided into four categories: gastropods (snails, limpets), bivalves (shellfish), crustaceans (shrimp, lobsters), and all other (worms, anemones). Apparently gastropods are the most resistant to the effects of oil; crustaceans are the most sensitive.

Most gastropods appear to have a rather high resistance to hydrocarbon toxicity; periwinkles, the common intertidal snail, apparently are very resistant. The critical concentration may be 100 to 200 ppm or more of soluble aromatic derivities. Limpets demonstrate the only significant deviation and appear to have a critical threshold concentration of less than 5 ppm. The relatively high resistance of gastropods may be due to secretion of a mucous substance which effectively shields them from the petroleum.

Bivalves, including oysters, clams, cockles and mussels, are moderately resistant to oil. The ability to close their shells and seal off the ambient water mass acts as an effective protection mechanism. However, this closed condition cannot be maintained indefinitely and, in fact, cockles tend to "gape", making them more susceptible. Typical critical concentrations for most bivalves are 5 to 50 ppm.

Both benthic crustaceans and certain other benthic organisms are apparently fairly sensitive to soluble aromatic derivitives. Threshold concentrations appear to be 1 to 10 ppm. Burrowing organisms may also be affected by alterations in the substrate texture and structure.

Carcinogens

The introduction and dispersion of cancer-causing compounds in benthic organisms is similar to that discussed for finfish. Carcinogenic compounds contained in crude oil vary; higher quantities are found in naphthenic and asphalt-based petroleum than in paraffin-based crudes. It should also be noted that the relative proportion of carcinogens per volume of crude oil will increase after weathering removes low boiling fractions. Hence, a spill of 10,000 tons of oil could contain 100 to 200 lb of carcinogenic material.

Carcinogenic compounds are exceedingly insoluble. Their solubility is increased by the presence of detergents or noncolloidal hydrocarbons, but the concentration of the detergents needed to achieve these increases is unrealistically high, especially in marine environment. Therefore, the primary mode of distribution of carcinogenic hydrocarbons is absorption (adherence) onto particulate matter. These carcinogens will tend to remain in the marine environment capable of being absorbed on particulate matter or taken up by burrowing animals, to provide routes to enter the food chain. Edible fish and shellfish can partially absorb these compounds through their gut tract. Marine animals do not appear to metabolize them to a significant degree when they enter their tissues. Potentially slow accumulation can occur; moreover, ample evidence is available to show this process does indeed occur for a number of fish and shellfish.

Tainting

Tainting of marine organisms also results from incorporation of hydrocarbons, causing the development of an objectionable taste. Bivalves and some finfish, such as mullet, are particularly susceptible to tainting. Crude oil concentrations in water as low as 0.01 ppm cause tainting in oysters. Lobsters become tainted only by immersion in seawater containing oil, but not from eating oil-coated food. This is particularly significant because it indicates that incorporation of hydrocarbons in the food chain may result from direct removal from seawater and not from feeding and ingestion of food containing petroleum substances. In the West Falmouth spill of Number 2 fuel oil, in a salt marsh, hydrocarbons were incorporated by nearly all organisms living in the marsh.

The concentration threshold for the development of objectionable taste in animal tissue is in the range of 5 to 50 ppm. Some data indicates that for various types of hydrocarbons, there may be maximum amounts that will be incorporated by a particular organism. For the mussel, it was found that aromatic hydrocarbon concentration in the organisms may reach 70 ppm (dry weight basis). Significant levels of hydrocarbons can be detected in the organisms within 2 to 4 hr after placement of mussels in oil contaminated seawater.

Archaeology and Historical Sites

These topics can be discussed in detail only with respect to a specific pipeline route through the marshes. However, in general, the number of historic sites in the marshes is sparse, and these are easily avoided so that the likelihood of impact upon a historic site would be most unusual.

There is a larger number of archaeological sites, believed to extend back nearly 2000 years. Although these are not too well documented, hundreds of sites are known. Most occupied sites were along natural levees where freshwater and dry land were available so that these tend to be located along present or past water courses. The impact from pipelines on the archaeology in the marshes and wetlands has generally been negative in the past. Apparently, some pipeline construction has gone through various Indian mounds without providing for any site examination. This is a matter which can be fairly easily changed by administrative procedures and by having some archaeological coordination of pipeline construction.

During an oil spill, any objects coated with oil would obviously be rendered less useful and valuable and may not survive cleaning operations. In addition, porous items, such as wood, pottery, or shell, may be internally contaminated with oil, and this might interfere with carbon dating procedures.

Esthetics

The impact on the esthetic qualities of the marshes is also generally considered to be negative. The primary reason for this feeling is the irreversible nature of the canals through the marshes. In a sense, it is like walking on the moon — the tracks never go away. The land-based observer, which will include most casual observers of the marsh, will not receive the full impact of the rearrangement of the marshes due to his limited horizon. The full impact is observed primarily by the aerial observer who can, in certain oil-field areas, see the marsh cut up for miles.

The relative isolation of the area, the lack of roads, and the general lack of external visitors has meant that to date any esthetic impacts have affected relatively small numbers of persons. However, any floating material, such as debris or oil, that is cast up on the beach or washed into a bay would constitute an impact upon the esthetic values for users or owners of the area.

Even after burial of the pipeline, the scars will cause an impact on the esthetic values of the beach and associated dunes and sand flats. It is estimated that the impact will endure for at least a year until sand has been redistributed by wind, tides, and rain and another growing season brings about revegetation.

Revegetation of dunes crossed by pipelines would reduce adverse effects from an esthetic and scenic viewpoint and would decrease the chance of destruction of the dunes by erosion. It is not, however, within the Federal Government's authority to require the revegetation of affected dunes unless they are on federal lands. State or local authorities may require revegetation of dunes disrupted by pipeline installations.

Impact on Recreation

Recreation activities in beach and shoreline areas can receive impacts from pipeline construction and from oil spills. As they cross a beach area used for recreation, there will be an impact. The area of beach disturbed by pipeline construction would be fairly small (about 40 ft wide), and the first high tides following burial of the pipeline would serve to restore the beach terrain. Restoration of the beach ridge would take longer, most likely requiring a storm tide or high winds to obliterate the effects of excavation. Physical interference, then, with recreational activities will be minimal and short lived.

Water sports, such as swimming, diving, spearfishing, underwater photography, fishing for finfish and shellfish, boating and water skiing, would be directly affected by an oil spill.

Other marine-related activities, such as beachcombing, shell collecting, painting, shoreline nature study, camping, and sunbathing, would be made much less attractive for an indeterminate period where an oil spill had coated a beach, camping area, or fishing site.

Removal of oil from beaches used for recreation in the area under consideration would probably involve removal of the contaminated sand and possibly, replacement of the sand if needed. The time period for clean-up in this case would depend on the extent of beach affected. Recreational use of the area would be precluded during the time that oil covered the beach and during the clean-up process also.

The impacts of an oil spill discussed above would be more keenly felt if the recreation area involved is intensively used or considered to have unique or outstanding recreational values such as many of the popular recreation areas along the eastern Gulf coast. Not only would the impact be felt by the recreational users of the area, but, consequently, the community or businesses whose economic well-being depends on use of their recreational resources by tourists would be affected. If an oil spill were to cover outstanding recreational areas during the height of the recreational season, the impact could be expected to be more severe, in that residents and tourists would not be attracted to a beach area contaminated by oil or undergoing a clean-up process, and there would be a resultant economic loss.

SUMMARY AND CONCLUSIONS

This chapter has attempted to define the particular environmental impacts that are likely to result from construction of pipeline rights-of-way in the coastal zone. The analysis of these impacts is no less important than studies undertaken to examine project hydrology, soil characteristics, or economic considerations that determine the desirability and cost effectiveness of a project. Too often these impacts are relegated to the intangible, without making a fair assessment of what these impacts are or will become, before choosing suitable mitigating adjustments. It is our belief that these impacts can be coped with in the same way that other factors can be dealt with in the determination of a project.

REFERENCES

1. Batelle Columbus Laboratories, Environmental Aspects of Gas Pipeline Operations in the Louisiana Coast Marsh, Final Report, Battelle Columbus Laboratories, Columbus, Ohio, 1972.
2. Bureau of Land Management, Outer Continental Shelf Oil and Gas General Lease Sale Offshore Louisiana, Environmental Impact Statement, Washington, D.C., 1973.
3. Bureau of Land Management, Outer Continental Shelf Oil and Gas General Lease Sale Offshore Mississippi, Alabama, and Florida, Environmental Impact Statement, Washington, D.C., 1973.
4. **Ortolano, L. and Hill, W. W.**, An Analysis of Environmental Statements for Corps of Engineers' Water Projects, Civil Engineering Department, Stanford University, California, 1971.
5. U.S. Army Corps of Engineers, Crude Oil and Natural Gas Production in Navigable Waters Along the Texas Coast, Environmental Impact Statement, Washington, D.C., 1972.
6. U.S. Army Corps of Engineers, Crude Oil and Natural Gas Production and Other Mining Operations in Navigable Waters Along the Louisiana Coast, Environmental Impact Statement, Washington, D.C., 1973.
7. U.S. Army Corps of Engineers, West Coast Deepwater Port Facility Study, Environmental Impact Statement, Washington, D.C., 1973.
8. U.S. Department of the Interior, Deepwater Ports, Environmental Impact Statement, Washington, D.C., 1973.
9. **Chapman, C.**, Marsh and Estuary Management Symposium, Louisiana State University, Baton Rouge, 1972.

Part II: Basic Nomenclature, Formulation And Application

Chapter 7

NOMENCLATURE AND MOLECULAR STRUCTURE

INTRODUCTION

Understanding herbicide chemical nomenclature and formulation is necessary to understand the proper use of herbicides. Since herbicides are chemicals, all of their properties and uses are dependent on their chemical structure, and similar structures have similar properties and uses. Practical application involves basic characteristics of herbicides as a part of the methodology of manufacture and formulation. Field application, likewise, is directly related to the chemical characteristics of the herbicides and the conditions under which they are applied to plants. The most exact name for herbicide is the chemical name. This tells exactly what chemical groups (functional groups) are contained in a herbicide and how these groups are attached together. With this knowledge, one knows exactly what the herbicide is if one can translate this name into chemical structure. Even without any knowledge of the herbicide except for the chemical name, one can estimate its properties and uses if one has the knowledge of the properties of the functional groups which make up the herbicide.

The second name for herbicide is called the common name. Chemical names (see Table 1) can be very long and cumbersome to use, whereas common names often give a hint as to which group the herbicide belongs. However, these names are not very descriptive of the application or use pattern.

The trade (commercial) name for a herbicide does not give information or understanding of the herbicide. Herbicides in use for a long time usually have many trade names, while the newer, patented herbicides usually have only one trade name (Table 2). The trade name is by far the most prominent; the chemical name can be found with some searching, and the common name is often omitted.

BASIC CHEMICAL NOMENCLATURE

The chemical name of a herbicide is simply the herbicide structure in written form. Hence, there is a direct relationship between the chemical name of a herbicide and its structure. Before we can concern ourselves with the properties and uses of herbicides, we must know their structures. Although some herbicides are inorganic, organic chemical herbicides make up a vast majority of the herbicide market.

The carbon (C) atom is one of the most important atoms, primarily because it has the unusual ability to attach itself to other atoms. Each C atom may attach to four other atoms, so we can say that each C atom has four valences. This is written as follows (R signifies any other atom or group of atoms):

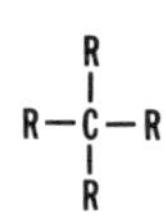

Special names are given when certain other atoms are attached to a carbon atom. When only hydrogen (H) atoms are attached to carbon atoms, the resulting compound is called a hydrocarbon as below (for simplicity, the hydrogen atoms are usually omitted):

Table 1
CHARACTERISTICS OF THE THREE DIFFERENT TYPES OF NAMES FOR A HERBICIDE

Name	Capitalization	Possible number of names in use	Used by	Difficulty of use	Relationship of words to structure
Chemical name	Never	1	Everyone (the cumbersome standard)	Very difficult	Absolute
Common name	Never (except when only a letter)	1	Everyone (the simple standard)	Easy	Often
Trade name	Always	Infinite	Salesman	Sometimes difficult (because of many names)	Almost never

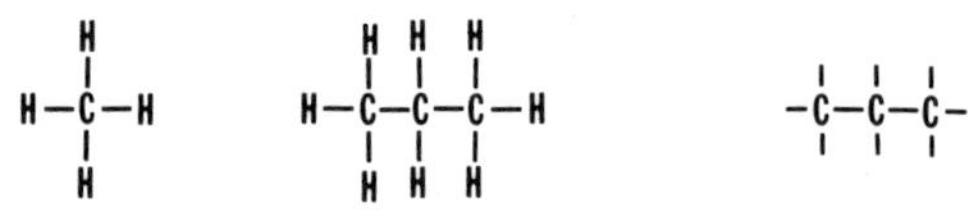

When oxygen (O) is attached (with double bonds) to a carbon atom, the resulting structure is called a carbonyl group.

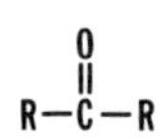

Several other groups which are commonly used in herbicides are

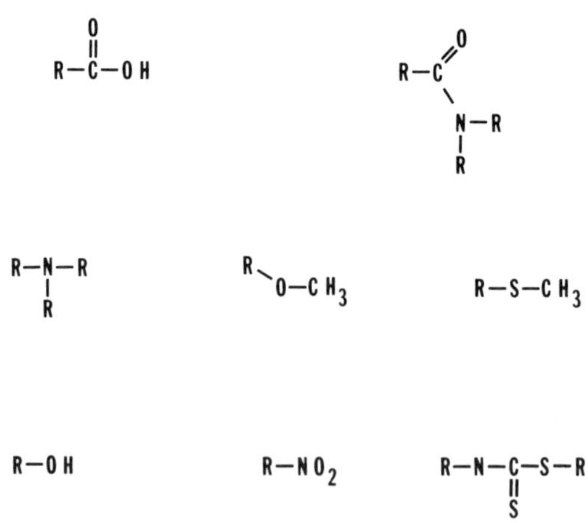

Table 2
EXAMPLES OF CHEMICAL, COMMON, AND TRADE NAMES FOR FOUR DIFFERENT HERBICIDES

Chemical name	Common name	Trade name(s)
2,4-Dichlorophenoxy acetic acid	2,4-D	2,4-D amine® Aqua-Kleen® Chem Pels® Crop Guard® Decamine®
2-*sec*-butyl-4, 6-dinitrophenol	dinoseb	Dow General® Chemox® Ivosit® Premerge® Sinox®
2-Chloro-4-(ethylamino)-6-(isopropylamino)-s-triazine	atrazine	AAtrex® 80 AAtrex 4L Bonus®, type S
S-ethyl dipropylthio-carbamate	EPTC	Eptam®

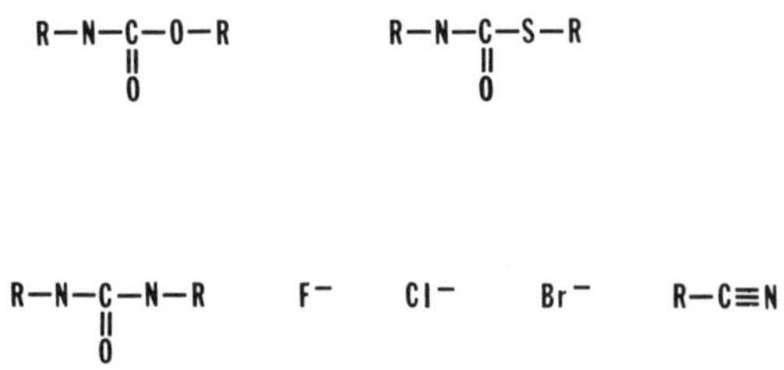

These groups can be attached to carbon chains of various lengths. Different names are given to different lengths of carbon chains as follows:

Number of carbons in a chain	Name of radicle (group)	Name of hydrocarbon	Name of acid
1	Methyl	Methane	Methanoic (formic)
2	Ethyl	Ethane	Ethanoic (acetic)
3	Propyl	Propane	Propanoic (propionic)
4	Butyl	Butane	Butanoic (butyric)
5	Pentyl	Pentane	Pentanoic (valeric)

Groups with greater than two carbon atoms have alternative configurations, e.g., butane:

Description	Structure	Nomenclature
Straight	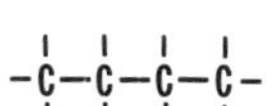	Normal (or *n*) butane
One chain	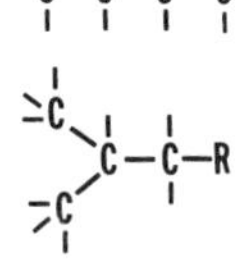	Isobutyl

Greatly branched chain

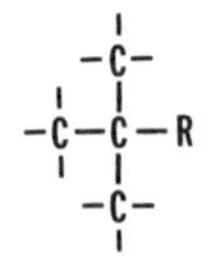

Tertiary (*tert*) butyl

HERBICIDE NOMENCLATURE

We may convert the chemical names of some herbicides into the corresponding chemical structures. One such example is the herbicide with the chemical name 2,2-dichloropropionic acid. It has two chlorine atoms (dichloro) both attached to the second carbon of propionic acid.

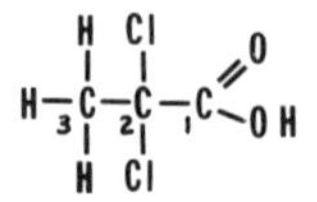

Carbon Number

Many herbicides have their C atoms attached in a circle (cyclic compounds) instead of in a straight or branched chain (aliphatic). The most common carbon cyclic compound is benzene which has six carbon atoms and six hydrogen atoms.

Carbons are numbered, as in the aliphatic compounds, starting from the largest group attached or from a nitrogen within the ring. In benzene and its derivatives, positions two and six are designated as ortho (*o*), positions three and five are meta (*m*), and position four is para (*p*). The radicle of benzene (with one H missing) is called the phenyl group. It combines with a hydroxyl group to form a compound called phenol.

DIFFERENT MOLECULAR FORMS OF THE SAME HERBICIDE

Several herbicides have as a part of their structure an atom which is not essential for their activity and is easily replaced by another atom or group of atoms. A new chemical group is, therefore, often added to the "parent molecule". This group is as easily broken off as it was easily attached. It affects herbicide properties before and very soon after application when the molecular reverts to its original form. Different forms of a herbicide can change certain properties of the parent molecule, such as solubility, volatility, and foliar penetration. Differences in herbicide selectivity, toxicity, and ease of handling can be regulated to some degree by making different forms of the same parent molecule.

The word "formulation" has been used loosely in herbicide terminology to include all additions to a parent herbicide whether these additions are actually chemically bonded to the molecule or whether they are just mixed into the same container. However, in this discussion, the molecular changes have been separated from the formulations (physical mixing of other substances with a herbicide). Except for gross differences in formulations, i.e., granular vs. liquid, this aspect of herbicides is ignored here because

1. It makes only slight differences in herbicide properties and uses.
2. Information is difficult to obtain.
3. There are hundreds of difficult-to-classify surfactants.
4. Formulations change continually.

Different molecular forms of herbicides do, however, make a rather significant change in herbicide properties and uses, and they are clearly defined and recorded on all herbicide labels.

Different molecular forms of a herbicide occur commonly when the parent molecule is an acid. The phenoxyalkane herbicides are most noted for differences in molecular forms. Amine and ester, for example, are words well known by people commonly using 2,4-D. Aliphatic acids and bipyridyliums are also commonly sold in molecular forms which consist of additions to the parent molecule. However, any herbicide can have an addition to its parent molecule if the chemist can perform the task economically and if a desired new property is obtained. For simplicity, the rates of application of different forms of a herbicide are calculated as if the parent molecule existed (ignoring the molecular change); this is why the term molecular equivalent is used. The term acid equivalent is usually used in a case where the parent molecule is an acid.

Additions of a chemical to a herbicide can be classified into three groups:

1. Salts (inorganic)
2. Amines (nitrogen organic salts)
3. Esters

For example, 2,4-D can be in its original acid form or in several salt, amine, or ester forms.

Different forms of (2,4-dichlorophenoxy) acetic acid follow:

1. Acid — slightly soluble in water (600 ppm).

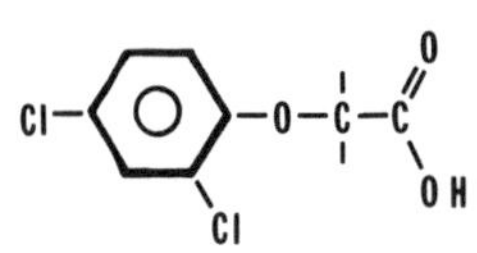

2. Salts — usually sodium (Na^+), potassium (K^+), or ammonium (NH_3^+) salts are used because they are more soluble in water than the acid. In hard water, magnesium (Mg^{++}) and calcium (Ca^{++}) salts are formed followed by precipitation because their salts are less soluble than the acid.

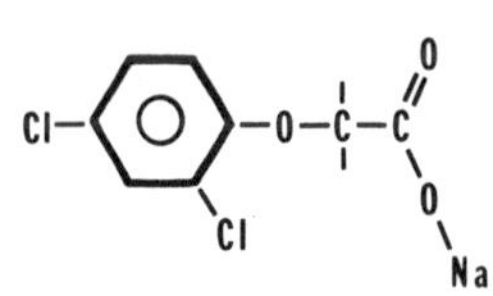

3. Amines — these forms are more water soluble than the inorganic salts and therefore easier to mix with water. They are the most commonly used form of the phenoxyalkane herbicides. Three common amine forms of 2,4-D are presented below.

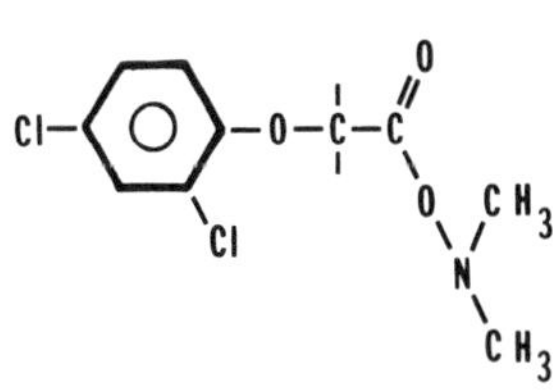

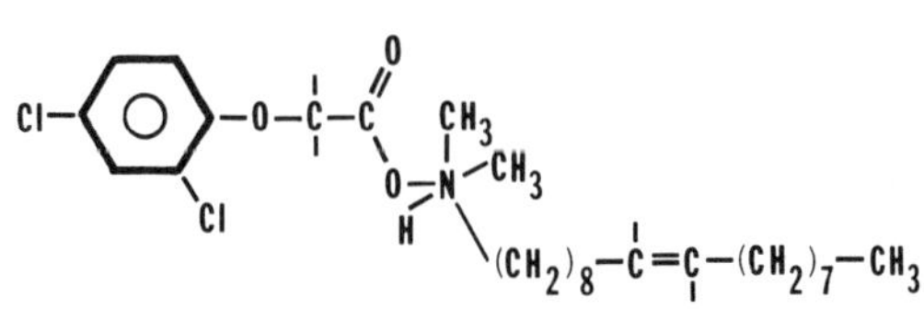

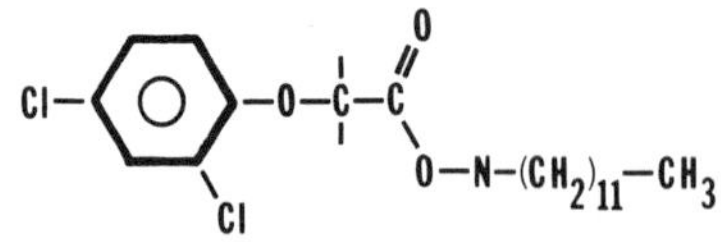

4. Esters — the ester forms are insoluble in water and soluble in petroleum oils. Although they are more volatile than the other forms of herbicides, the degree of volatility depends on the type of ester. In general, the longer the side chain, the less volatile. The esters with long side chains are often referred to collectively as low-volatile (LV) esters. However, volatility of LV esters can still be important under hot weather conditions. High volatility and low polarity which increases penetration into the plant makes esters more toxic to plants than do the other forms.

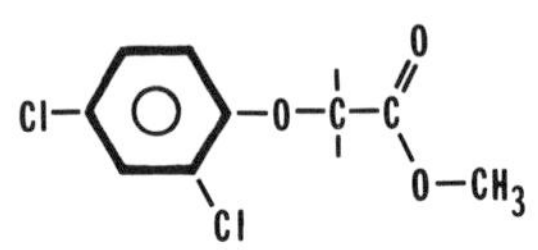

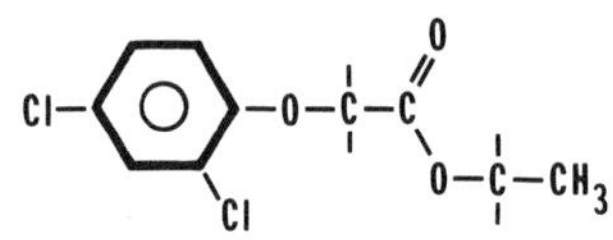

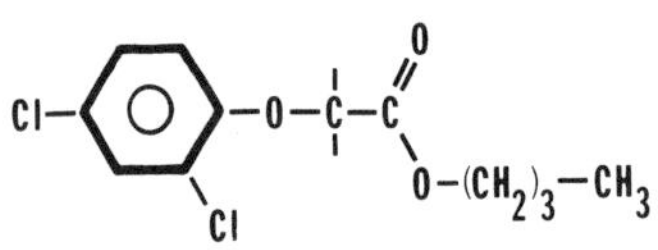

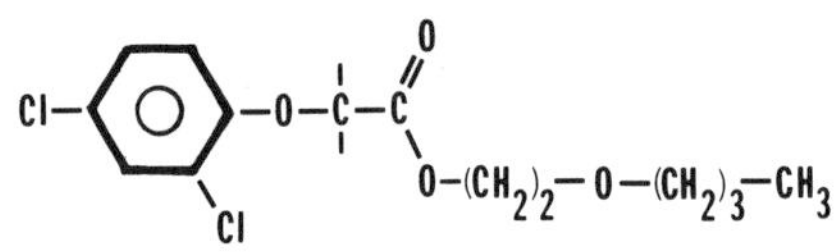

REFERENCES

1. Ahlgren, G. H., Klingman, G. C., and Wolf, D. E., *Principles of Weed Control,* John Wiley & Sons, New York, 1951.
2. Crafts, A. S. and Harvey, W. A., Weed control, *Adv. Agron.*, 1, 289, 1949.
3. Department of the Air Force, Land Management and Grounds Maintenance, Air Force Manual 85-6, U.S. Government Printing Office, Washington, D.C., 1963.
4. Department of the Army, Grounds Maintenance and Management, Technical Manual TM 5-630, U.S. Government Printing Office, Washington, D.C., 1950.
5. Department of the Navy, Maintenance of Grounds, NAVKOCKS MD-100, U.S. Government Printing Office, Washington, D.C., 1963.
6. Department of the Army, Herbicide Manual for Noncropland Weeds, Technical Manual TM 5-629, U.S. Government Printing Office, Washington, D.C., 1970.
7. EPA, Acceptable Common Names for the Ingredient Statements on Pesticide Lables, EPA 540/9-75-Oil, U.S. Environmental Protection Agency, Washington, D.C., 1975.

Chapter 8

FORMULATION AND HANDLING OF HERBICIDES

INTRODUCTION

A formulation contains the herbicide in a form that can be

1. Dissolved or suspended in a carrier and distributed in solution or suspension by sprayers
2. Distributed dry by dusters or spreaders
3. Easily vaporized for fumigation

An emulsifier, spreader, sticker, or other surfacant is often added to facilitate dilution or increase wetting capacity. Frequently, formulations contain an inactive filler that serves as a diluent only. For example, there may be only 2, 3, 4, or 6 lb of active weedkiller in an herbicide formulation that weighs 10 lb/gal. The following define several types of formulations.[1]

FORMULATION

Water-Soluble Concentrates

Water-soluble concentrates are readily dissolved in water to make a spray and are often sold as solids. Trichloroacetic acid (TCA), dalapon, sodium, salt of 2,4-dichlorophenoxy (2,4-D), and amitrole are examples. They are also formulated as liquids, i.e., amine salts of 2,4-D and the sodium salts of PCP and endothall. These liquid formulations are prepared because some water-soluble chemicals cannot be handled conveniently unless they are diluted. For example, the alkanol amine salt of 2,4-D is soluble in water, but at room temperature, it is a heavy viscous liquid that would be difficult to measure or mix with water, and when cool it becomes stiff and unpourable. It is formulated into a free-flowing liquid that easily can be diluted with water and is convenient to handle. Water-soluble concentrates are also available as pastes and slurries — sodium salts of pentachlorophenol (PCP) and 2,4-D. They may or may not include wetting agents.

Emulsifiable Concentrates

Emulsifiable concentrates are usually liquids in which the chemical is dissolved in one or more water-insoluble solvents, such as oil or benzene, and to which an emulsifier is added. When the emulsifiable concentrate is added to water and agitated, the mixture breaks into fine droplets. The emulsifying agent causes the fine droplets of oil to be suspended in the water to form an emulsion. The esters of 2,4-D, 2,4,5-trichlorophenoxy (2,4,5-T), and silvex are samples. The formulation should be clear, homogeneous, and free from sediment or crystalline solids. It should not corrode equipment, and the flash point should not be lower than 140°F (60°C). Since emulsions are frequently used in cold weather, the concentrate should have a maximum pour point of −10°F (−23.3°C).

The emulsifiable concentrate should disperse readily. A concentrate having poor dispersibility emulsifies only with vigorous agitation, and free oil usually separates from the emulsion on standing a short time. A good emulsion has globules that are barely visible; one with globules so fine they cannot be seen with the naked eye is better. Such an emulsion is called tight.

The breaking time of an emulsion is also important. Some formulations produce emulsions that break in a few minutes after application; others remain stable for 24 hr. The fast-breaking emulsion requires constant agitation and does not wet leaves adequately. Where constant agitation is impracticable, such an emulsion settles in the tank. Very stable emulsions may not break soon enough, expecially in brush spraying, and they tend to drain off the leaf. A good emulsion should contain the solvents and emulsifying agents necessary to keep it stable with hard or soft water during the mixing and application period, but allow it to break soon after reaching the plant surface. As an average, the hardness of water is 322 ppm in terms of calcium carbonate for hard water, 115 ppm for water in municipal systems, and 35 ppm for rainwater. Emulsifiable concentrates should be stable under both hot and cold storage. Invert emulsions are water-in-oil mixtures in which every spray droplet is surrounded by oil instead of water. This results in a viscous material difficult to apply but less likely to drift.

Oil-Soluble Concentrates

Oil-soluble concentrates are similar to emulsifiable concentrates, but do not contain an emulsifier. They do not mix with water, but they can be diluted with fuel oil or kerosene, or applied without dilution (e.g., PCP and dinoseb).

Powders and Pastes

Wettable powders or water-dispersible powders are generally formulated from chemicals that are insoluble or soluble with difficulty in oil or water. They are finely ground with or without a powdered diluent. A wetting agent is added to keep the particles from floating, and a dispersing agent is added to keep the material in suspension. Monuron, diuron, and simazine are examples; 2,4-D and 2,4,5-T have been formulated as powders and pastes.

Granules and Pellets

Granules and pellets are forms of herbicides in which the chemical is impregnated upon vermiculite, attaclay, or similar carriers. They can be prepared by spraying the herbicide on the carrier or by adding the core material, such as vermiculite, to the original dry powder. Then a pillrolling technique with binders is used to produce pellets, or the powder plus binders and water are extruded, chopped, and dried to produce grains or pellets. Fenuron, simazine, 2,4-D, and an increasing number of herbicides are formulated as pellets; no water or mixing is required, application equipment is less expensive than sprayers, and they can be applied in areas awkward for spray equipment. If the chemical is effective only by root absorption, a spray intercepted by leaves is wasteful; grains or pellets that drop to the ground are more effective. However, their cost is higher, they are not adapted for foliar applications, but must be applied to the soil, they are generally less effective than sprays where moisture is limited and distribution of small amounts is difficult except by hand.

Adjuvants and Surface-Active Surfactant Agents

Adjuvants are substances added to an herbicide or spray to improve its action. They may be added by the company at the time of the formulation of the active ingredient or at the time of the mixing of the spray solution before application. Such substances may enhance toxicity, assist emulsification, increase spreading properties, promote leaf retention and pentration, or perform other functions. Certain adjuvants reduce interfacial tension and are known as surface-active agents or surfactants. Surfactants are adjuvants that bind two or more incompatible phases — such as water and oil — in more intimate contact by modifying the interfacial forces between them.

There are various types of adjuvants and surfactant agents. Wetting agents reduce the interfacial tension and bring a liquid into intimate contact with a solid, for example, increasing the spread of a contact spray on the leaf surface. Spreading agents are wetting agents that improve spreading properties of water on leaf surfaces. Emulsifiers are used to maintain the stability of an emulsion. Relative apolar and polar portions of the molecular have affinitives for oil and water, respectively. Surfactant molecules orient at the interface between oil and water surfaces and couple them together, and as a result, tiny dispersed droplets in a stable emulsion are prevented from coalescing. Thickening and sticking agents increase spray adherence to leaves and reduce bounce or runoff during spraying. Penetration agents are substances that assist plant absorption of an herbicide. Dispersing agents reduce cohesion between like particles. They play the same role with the solid organic portion of a wettable powder in water that an emulsifier plays in a liquid system.

On many difficult-to-kill weed species, the use of adjuvants may greatly increase the killing action of most herbicides. However, adjuvants may reduce selective action.

PURCHASE OF HERBICIDES

Adequate specifications for some herbicide chemicals have been developed. Basically, the economic value of an herbicide depends upon the relative amounts of active chemical toxic to plants that are contained per pound or gallon. This is expressed in percent of active ingredient, acid or phenol equivalent for solids, and pounds per gallon for liquids. Thus, a solid containing 74% of the weedkilling ingredient is worth more than a 20% granular product pound for pound, and an amine salt formulation of 2,4-D containing 4 lb/gal acid equivalent is of more value than the same formulation containing 2 lb/gal. Actually, since mixing charges and cost of containers, freight, and handling have to be paid on twice as much material for a 2-lb/gal formulation as for a 4-lb formulation, the cost per pound of active material is less in the 4-lb formulation than in the 2-lb formulation, although the cost per gallon of product is more.

However, this relationship is affected by other factors. Under some circumstances, 1 lb of 2,4-D in the ester formulation is twice as effective as 1 lb of the amine salt. This could be true on hard-to-kill species, on many brush species, and in arid climates. Equivalent amounts of the sodium salt are less effective than the amine salts. The amount of chemical that can be "packed" into a formulation is limited both for physical, chemical, and economic reasons.

Some herbicides are mixtures of several isomers. Often one is more effective than the others. Although it may not be economically practicable to purify a formulation for the effective isomer, there may be larger proportions of it in one formulation than in another. For example, among the polychlorobenzoics, the 2,3,6-trichlorobenzoic (2,3,6-TBA) acid is more effective on most species than the 2,3,4- , 2,3,5- , 2,4,5- , 2,4,6- , or 3,4,5-TBAs.

Additives, such as wetting agents, spread-stickers, cosolvents, and emulsifiers, influence the effectiveness of a chemical. Their precise role and comparative values, however, are not too well understood, and purchase at present must depend largely on empirical information. Carriers of the chemical also vary in cost and in toxicity to plants; some can reduce the effectiveness of the chemical.

The proprietary mixture of two or more chemicals may be less effective than the constituents applied separately. Because better timing of application can be made of each constituent applied separately than applied together, the effectiveness of the mixture may be less than the sum of individual effects.

Currently, herbicide formulation purchases depend on the relative amounts of active chemical per unit, packaging, freedom from foaming and precipitation problems, and stability under storage. Sometimes the superiority of a formulation can be demonstrated because of these factors. When federal specifications are available for all herbicides, purchases can be made on the basis of the requirements stated in the specifications.

MIXING OF SPRAY MATERIALS

Mixing spray materials correctly involves calculations of the proper amounts of herbicide and diluent and procedures for mixing.

Calculations

Rates of application are expressed in terms of acid equivalent for certain herbicides (such as 2,4-D, 2,4,5-T, 2,3,6-TBA, dalapon, and Silvex®), phenol equivalent for dinoseb, and active ingredient for others (such as amitrole and simazine). They are recommended in ounces or pounds per square foot, square rod, or acre, where the area to be treated can be measured.

When spot spraying of individual or small clumps of plants is necessary, or where dense foliage (as in brush or trees) is to be thoroughly wet, the concentration of the spray is recommended in pounds of active ingredient or acid equivalent per 100 gal. Also, in the control of aquatic weeds, concentration of the herbicide is often referred to in terms of parts per million of water, i.e., the number of parts (by weight or volume) of the chemical in 1 million parts of water or other diluent.

To determine the volume of water in a pond and the amount of chemical needed, the following formula is used:

1. Average depth of water (in feet) × surface acreage × 43,560 = volume water (in cubic feet) or average depth of water (in feet) × surface area, in square feet = volume water (in cubic feet)
2. 1 acre-ft of water weighs 2,722,500 lb; i.e., 43,560 ft^2 × 1 ft = 43,560 ft^3 × 62 lb = 2,722,500 lg
3. Therefore, 1 ppm = 2.7 lb chemical in 2,722,500 lb water (1 acre-ft)

To determine number of pounds chemical required, use the following: X (pounds chemical) = parts per million × acre-foot × 2.7.

For example, if the rate is 5 ppm, volume is 1 acre 2 ft deep, then X = 5 × 2 × 2.7 = 27. Therefore, 27 lb of chemical would be required for 5 ppm × 2 acre-ft. Usually, the pounds of active ingredient or acid equivalent per gallon are given on the label of liquid herbicide. The percent is given on labels of powders, granules, and other dry materials.

To calculate the amount of liquid herbicide required when the rate is expressed in pounds per acre, use the following formula:

$$\frac{\text{rate in pounds per acre}}{\text{pounds of herbicide per gallon}} = \text{gallons per acre}$$

For example, if the rate is 1 1/2 lb/acre and the herbicide contains 4 lg/gal, 1.5 = 0.37 gal, or 8 pt × 0.37 = 3 pt.

Use the same formula to calculate gallons of herbicide per 100 gal of spray.

For example if the rate is 2.5 lb/100 gal and the herbicide contains 2 lb/gal, 2.5/2.0 = 1.25 of 1 1/4 gal.

To calculate the amount of dry product required when the rate per acre is given, use the following:

$$\frac{100}{\text{percent active ingredient}} \times \text{rate per acre} = \text{pounds product}$$

For example, if the rate is 15 lb active ingredient per acre, and the percent of active ingredient is 75, then $100/75 \times 15 = 20$.

Tank Capacity

To calculate capacity of the sprayer tank (all measurements in inches), use the following:

1. Cylindrical tanks: length × square of diameter × 0.0034 = capacity in gallons
2. Tanks with elliptical cross section: length × short diameter × longer diameter × 0.0034 = capacity in gallons
3. Rectangular tanks with square or oblong cross section: length × width × depth × 0.004329 = capacity in gallons

If tank capacity is not known, it can be determined in a practical way by filling the tank with water and then emptying it into a measure of known capacity, such as 1- or 5-gal container.

Never pour the concentrate into an empty tank. Either fill the tank half full, add the chemical, agitate, and complete the filling or start filling and add the chemical as the filling is continued. Operate the sprayer with the nozzles shut off, by-passing the spray through the tank for several minutes (at least six times) to insure a thorough mixing.

Water Sprays

When using a soluble powder or crystals, fill the tank half full of water, add the chemical slowly, and stir to dissolve it completely. If a detergent is recommended, add to ½ lb of household detergent per 100 gal of spray. If a solid grasskiller is to be used with 2,4-D or 2,4,5-T, mix it with water first and then add the liquid herbicide. Use the spray solution within 1 or 2 days after mixing, since some chemicals lose strength on standing.

If an amine salt of dinoseb is used in very hard water, add 1 lb of softener per 100 gal of spray.

Emulsions

Since emulsions are suspensions and not solutions, they require constant agitation. If no additional oil is to be used, start running water into the empty tank and slowly add the herbicide formulation with continuous agitation. The chemical should all be added by the time the tank is one third full.

If oil is to be used, premix the herbicide and the oil in a separate container. Do not allow any water to get into this mixture, or it may jell. Fill the tank one third full of water, add premixed oil and herbicide, and continue filling. Agitate constantly. The water should be clean and low in carbonates and sulfates. (If ordinary hand soap lathers well, the water is satisfactory.) Circulate the mixture until it is uniformly white. If the emulsion stands for several hours, stir until uniformly white before using. Take samples of the emulsion from various sections of the tank at intervals to determine if separation has taken place.

When an invert emulsion is used, add water slowly to the premixed herbicide-oil solution under constant mechanical agitation; circulating with by-pass is insufficient. Invert emulsions may separate readily on standing. After 24 hr of separation, most invert emulsions cannot be restored.

Oil Sprays

Add esters of 2,4-D, 2,4,5-T, or Silvex® or other oil-soluble herbicides to the required amount of oil, and mix thoroughly in the tank. This can be done at any time before spraying, since the spray does not separate. Do not let any water or oil-water mixture get into the chemical formulation or the herbicide-oil mixture, or it may jell.

Suspensions

Water-dispersible powders mix readily with water, but do not dissolve. There are two methods of mixing: (1) add the powder slowly to water, with agitation (it should not be added to an empty tank), or (2) make a thin water slurry of the required amount of material before it is added to the water in the tank. A thorough job of mixing will help the agitator do its job more efficiently.

When esters of 2,4-D are used with monuron or diuron, add 1/2 lb of a caseinate conditioning agent to the suspension before adding the 2,4-D ester. When a water-soluble chemical is to be mixed with a water-dispersible powder, dissolve the water-soluble chemical in the water before adding the powder.

STORAGE OF HERBICIDES

A chemical with a low flash point (4.4°C or less) is dangerous in storage. Deterioration of chemicals in storage can be prevented by observing the following precautions.

Precautions

Dusts and wettable powders present no problem as long as they are kept clean and dry. They cake when wet, and packages may deteriorate. Water-soluble solids also cake when wet and when subjected to great changes in temperature. If packages are left open, hygroscopic chemicals become wet by absorbing water from moist air. Such chemicals as TCA, dalapon, PCP, chlorates, borates, and ammonium sulfamate cake when they are wet.

Liquid formulations should be stored on pallets or duckboards to keep the metal containers from rusting. The containers should be tightly closed, and air vents punched in cans to facilitate pouring should be plugged. Even small amounts of water introduced into emulsion concentrates or oil solutions can make them jell and cause deterioration of the container.

Chemicals may crystallize out of solution at temperatures below 0°C. If this happens, warm the products to 4.4°C or higher and roll the drums or shake the containers. If the crystals return to solution, no harm has been done. The following formulations do not freeze if stored in unheated rooms: low-volatile esters of 2,4-D, 2,4,5-T, 2,4-D plus 2,4,5-T (brush killer), and Silvex®.

The following formulations may crystalize out of the solution if stored at temperatures below those indicated:

Formulation	Storage temperature (°C)
2,4-D, alkanol amine salt	−9
2,4-D, dimethylamine salt	−7
2,4-D, butyl ester	−4
2,4-D, isopropyl ester	0
2,4,5-T, butyl ester	8
2,4,5-T, amine salt	0
2,4-D plus 2,4,5-T, butyl ester	−10
2,4-D plus 2,4,5-T, amine salt	6
MCPA, amine salt	−9

At high temperatures, chemicals may expand and cause bulging of drumheads and leaks in the containers. They may also deteriorate at temperatures over 35°C. High temperatures may reduce the effectiveness of emulsifiers and hasten the corrosion of containers.

REFERENCES

1. Farm Chemicals, *Pesticide Dictionary,* Meister, Willougby, Ohio, 1975.
2. Valkenburg, W. V., Ed., *Pesticide Formulation,* Marcel-Dekker, New York, 1973.
3. Ashton, F. M. and Crafts, A. S., *Mode of Action of Herbicides,* Interscience, New York, 1975.
4. Klingman, G. C. and Ashton, F. M., *Weed Science; Principles and Practices,* Interscience, New York, 1975.

Chapter 9

APPLICATION EQUIPMENT AND OPERATION

INTRODUCTION

The results that follow the use of a herbicide depend largely on how well or how poorly it is applied, and this, in turn, depends on the operator and his equipment. Equipment is designed to apply herbicides

1. Wet, in sprays or mists
2. Dry, in dusts or granules
3. In readily vaporized form for fumigation

EQUIPMENT

Hand Sprayers

Hand sprayers are suitable for treating small patches of inaccessible areas and fence rows, spot spraying, and for situations where the spray is to be applied close to susceptible plants. They include sprinkling cans, compressed-air sprayers of 1- to 5-gal capacity carried by a strap over the shoulder, and knapsack sprayers with hand-operated pumps. In the ordinary compressed-air sprayer, the pressure is reduced as the tank is emptied, however, there are also constant-pressure types. Some models, useful on trails, have cylinders of CO_2 gas to provide pressure. Knapsack sprayers are more expensive than compressed-air sprayers; however, they have an agitator attached to the pump that makes them more suitable for suspensions.

Water-dispersible powders that form suspensions of high concentration in water tend to clog flat spray-tip nozzles; a cone type gives less trouble. Compressed-air and knapsack sprayers are well adapted for herbicide concentrates to be used along highways. The cone oil-burner type nozzle that delivers 2.5 to 6 gal/hr at 20 to 30 lb/in.^{2} is very suitable. The solution or emulsion at 20 times usual concentration is applied at the rate of 5 to 6 gal/acre for low-growing vegetation or 12 gal/acre for heavier brush. A hood around the wand or boom prevents drift onto sensitive plants nearby.

Boomless Sprayers

Boomless sprayers — nozzle-cluster type — are well adapted for spraying roadsides and ditchbanks, under utility lines, and along fence rows. They are less expensive, simpler to operate, and have less nozzle trouble than boom sprayers. They can pass between trees and shrubbery, be maneuvered close to other obstacles, and are practical for rough ground. This type sprays a broadcast swath of 20 to 30 ft with large volumes that provide moderately good coverage.

For foliage sprays, the conventional equipment consists of a piston-type hydraulic sprayer (such as the Hardie, Bean, Iron Age, or Myers) equipped with 3/4-in. high-pressure hose and 15 gpm (gallon per minute) pumps for a 3/16-in. nozzle opening, or 35 to 60 gpm pumps for 1/4-, 5/16-, and 15/16-in. nozzle openings.

The equipment is mounted on a four-wheel drive power wagon or caterpillar-type tractor. The chief disadvantage of the boomless sprayers is that the spray stream is greatly affected by the wind. They should not be used when the wind may cause drift to sensitive vegetation nearby.

Boom Sprayers

Boom sprayers are adapted for either large areas where complete coverage is necessary or turf areas adjacent to a road where the entire application can be made from a tractor or truck-mounted boom.

Arm booms are available for roadside or ditchbank spraying. One end of a shaft is fastened to the truck, and the other end terminates in a boomless sprayer nozzle. A two-nozzle arm boom is used for ditchbanks. A larger boom can be set at right angles to the truck for roadside or ditch spraying or can be set vertically to spray above tall weeds. For roadside spraying, two or more nozzles are grouped together and mounted on an arm that reaches over mailboxes, highway signs, etc. A truck-mounted boom designed to spray under guide rails reaches over these obstacles and sprays from the outside toward the pavement. Nozzles mounted off-center enable the operator to spray 15 to 20 ft on one side.

Hand booms and adjustable spray guns are operated from truck-mounted power spray equipment and are used for application of basal sprays and stump equipment and are used for application of basal sprays and stump treatments, brush control and spot treatments, and for spraying around structures of various kinds. Handguns are used with pressures of about 100 $lb/in.^2$. Increased pressures result in a larger proportion of fine droplets, and drift is increased. Handguns can vary delivery from a broad mist to a narrow coarse stream. Instead of a single nozzle, a bank of three to eight can be used. Sometimes the conveyor truck is stationary, and long lines of hoses and side hoses are used to reach the area to be treated.

Mist Blowers

Mist blowers are power-driven machines that disperse highly concentrated sprays in finely atomized form at low volumes per acre. The herbicide is carried principally in an airstream instead of a liquid. These sprayers are free from boom and nozzle troubles and very useful in spraying herbaceous weeds and woody plants in swamps, under fences, around stone piles, along roadsides, in drainage ditches, and under powerlines. Mist blowers require minimum amounts of water, cover vegetation rapidly, and can be used for areas inaccessible to power equipment. The equipment is cheaper than the conventional hydraulic power equipment. However, use of mist blowers for weed spraying is limited by the serious hazard of drift. Mist blowers in 5- to 12-hp sizes are useful for brush spraying, if drift is not a problem. A 2-hp knapsack mist blower is useful for brush up to 30 ft tall, spot spraying, and for retreating.

Special Equipment

A right-hand drive truck with four nozzles mounted on a 6-ft sturdy vertical beam is often used for spraying berm and adjacent weeds and brush. A low-position, solid-stream nozzle with a 2/15-in. opening sprays berm to 24 ft from the machine, and three fan-type boomless nozzles spray brush and weeds close to the machine. The pump delivers up to 20 gpm at 800 $lb/in.^2$ from a three-speed power takeoff. The whole operation is controlled by the driver of the truck. The sprayer delivers from 25 to 50 gal semiconcentrated material per mile of road.

For spraying around guardrails and posts, there is the truck-mounted tank with power sprayers and a boom mounted on the right side of the truck. The boom is controlled by an operator sitting next to the driver. The control valve is handoperated. There are two nozzles; each sprays on either side of the guardrail or post. A boom operated hydraulically or electrically makes it possible to raise or lower it over obstructions, posts, and slopes for treating shoulders or fireguards. The boom for shoulder applications is about 8 ft long; for fireguards, it is 12 ft long.

For basal-stem treatment, a 4-hp gas engine and 1/2-in. gear pump and two lines of light hose with suitable nozzles, gun, and shutoff valve attached to each line of hose can be used. The spray is five times the concentration of dilute sprays. A simple device is a 2-gal can called a "trickler" which is carried on the operator's back and allows the chemical to flow by gravity through a flexible tube to a light wand. The discharge is controlled by a leakproof valve.

For bush, briers, and small trees in fence lines and on ditchbanks, there is the tractor-powered, high-pressure sprayer for spraying foliage. It may be equipped with a nylon roller or other high-pressure pump, 25 ft of high-pressure hose, and an adjustable spray pattern handgun.

Individual Parts

Pumps

Pump capacity should be about twice the nozzle delivery rate to provide for an overflow that is by-passed back to the tank for agitation of the spray. The several types are discussed below:

1. A rubber-impeller pump is cheap, not injured by abrasive suspensions, develops about 35 $lb/in.^2$, and is not adapted for oils.
2. A rotary-gear pump has positive action. The discharge rate depends on the speed of rotations. Suspensions are hard on the gears. Plastic gear pumps can be replaced more cheaply than repairing permanent equipment.
3. A centrifugal pump is a single-rotating impeller type. It pumps a large volume of spray, but does not develop a high pressure — 40 to 70 $lb/in.^2$ It can handle all spray materials with minimum wear. It is not self-priming, so it must be mounted lower than the tank. Unlike most pumps, it pushes the liquid in one direction only.
4. A piston pump is designed for large quantities of spray and high pressures — up to 1000 $lb/in.^2$ It has one or more plungers connected to a crankshaft. The plungers or pistons work inside smooth cylinders. The piston pump can be used for any type of sprayer, stands rough treatment, and is long lasting, but expensive.
5. A nylon-roller pump gives good service, but it is expensive and not well adapted for suspensions. It pumps only after reaching a highhpeed.
6. A diaphragm pump is similar to a piston pump except that one side of the chamber is made of a flexible fabric that creates the vacuum. This pump handles abrasives well, but the fabric does not always last long under the pressure normally required for spraying. The diaphragm, however, can be replaced easily and economically.

Tanks

Tanks should be easy to clean and rustproof. Other desirable features are a wide mouth with 12- to 16-mesh removable strainer; a round or trough-shaped bottom with a drain in the lowest part, a paddle-type agitator that moves slowly, sweeps the bottom, and can be disengaged, and an auxiliary 1- to 3-gal graduated flush tank for quick flushing of the spray line which can also be used for small quantities of spray. The tank should be calibrated in 5-gal increments with an unpainted and unvarnished measuring stick. A shutoff valve should be placed immediately below the tank to prevent spray material from settling in the spray line, strainers, and nozzles.

Pressure Regulators

By-pass regulators are necessary for gear, nylon-roller, diaphragm, and reciprocating piston pumps; diaphragm regulators are needed for centrifugal or impeller pumps. The by-pass regulator allows excess liquid to flow back to the tank. With proper adjustment, it can maintain a constant pressure at the nozzles. Neither a spring-loaded ball-type high-pressure relief valve nor a handoperator by-pass valve functions correctly as a pressure regulator. The regulator should be a sensitive spring-loaded type for high pressures and a disk or diaphragm type for pressures up to 175 $lb/in.^2$

Valves, Pressure Gages, and Strainers

The shutoff valves should be fast, easily accessible, and of the same diameter as the main boom line. Dripping may be reduced or prevented by special spring-activated valves installed between each nozzle and the boom that automatically close with reduced pressure or by venturi tubes and shutoff valves, arranged to cause a suction from the nozzles into the boom when the supply is cut off. Pressure gages should be mounted where they can be easily seen by the operator and should be large enough to be easily read.

Strainers in the boom line should be placed so as to intercept solid particles before they reach the pump, pressure regulator, or nozzle. A 50- to 80-mesh screen should be installed in the suction line to the pump. A strainer of 100- to 150-mesh and about 100 $in.^2$ in area should be placed between pressure regulator and boom when emulsions or solutions are used. Nozzles require individual screens — a 50-mesh for tips of greater capacity than 0.1 gal/min and 100-mesh for smaller tips. Nozzle screens should have mesh openings only slightly smaller than the nozzle opening, so that fine particles may pass through.

Hoses and Fittings

Hoses should be strong enough to withstand the pressure to be used and of a material not deteriorated by spray solutions and oil; ordinary garden hose stands about 70 $lb/in.^2$ The hose on the intake side of the pump should be at least 1/2 in. in diameter. The discharge hose from pump to pressure regulator should be of the same size. The discharge hose from the pressure regulator to the nozzles can be smaller, but must be large enough to feed at least twice the number of nozzles being used. Too small a hose results in reduced pressure at the distant nozzles. Pipe fittings should be resistant to corrosion. Eyelet fittings are more desirable than welded.

Booms

Booms should be adjustable for height, and the tube that carries the spray should be noncorrosive. In booms with large diameters, the spray liquid does not flow as rapidly as in those with smaller diameter. When a suspension is used, the powder may settle in the boom if the flow is too slow. Valves at the outer ends facilitate cleaning. Dry booms are supports for a separate hose or tube that carries the spray liquid to the nozzles. Small booms can be used for hand sprayers instead of a wand. A funnel slipped over the nozzle on the end of a wand confines the spray to a small area and prevents contact with nearby plants.

Nozzles

Nozzles are of two general types, those with convenient removable tips and those whose tips are an integral part of the nozzle. Tips are designed to produce a flat, fan-shaped spray or a cone-shaped spray. The fan-shaped nozzle is the more accurate in volume delivery and produces a spray that covers more uniformly than the hollow-cone type on the average boom sprayer. The fan-type nozzle with tapered edges sprays more uniformly on a boom than the preemergence or band-spray type with square edges. Cone-shaped nozzles do not clog as easily as flat-spray types, especially with the spray is a suspension. Broadcast nozzles spray a wide swath. They are used singly and in clusters to cover a width of 20 to 30 ft and, under some conditions, up to 60 ft. Uniform spray patterns are difficult to obtain, especially if there is a wind.

The most widely used fan-type nozzles are the flat atomizing group of which Tee-Jet is an example. They come in spray angles of 110 to 15° or even in a solid stream. Spacing on a boom is usually 20 in. On the nozzles of some companies, the first two digits of the number on a flat-spray tip indicate the spray angle, the latter two indicate the capacity in gallons per minute at 40 $lb/in.^2$. Numbers on cone-spray tips designate the capacity only at 10- , 40- , or 60-lb pressure and depend on the tip design. The spray angle is directly associated with design and the line pressure and is stated in the manufacturer's specifications.

Flooding types of flat-fan nozzles deliver a spray whose droplets are about one half larger in diameter than those delivered by the conventional flat-fan nozzles and produce a spray pattern of larger droplets than those produced by the hollow- or full-cone types.

The smaller the opening, the finer the droplets provided the nozzle design and pressure are the same. Larger openings can be used for any method or application by:

1. Widening the nozzle spacing on the boom
2. Increasing the spray volume per acre
3. Increasing the rate of travel
4. Decreasing the pressure

Flat-spray off-center 3/4-in. tips should be used for ditchbank weeds, while 1/4-in. off-center tips should be used for directed basal sprays. Two small fan-type tips in a double-swiveled nozzle are best for hard-to-wet foliage instead of one larger-sized tip, e.g., two 8002s instead of one 8004.

Aerial Sprayers

The obvious advantages of aerial spraying are offset to some degree by the hazards to the pilot and the damage from drift. A summary of pertinent information follows.

A helicopter needs no runway and can fly better in low visibility than light fixed-wind aircraft, but is more expensive. Drift can be reduced by

1. Flying in calm weather
2. Using low-volatile chemicals
3. Using nozzles with large openings
4. Using water or emulsions instead of pure oils
5. Using low pressures

A positive shutoff valve for each nozzle will avoid damage from dribble.

Tanks — Tanks are made of stainless steel, aluminum, galvanized iron, or molded plastic. Removable tank liners of synthetic rubber or plastic shorten cleaning time and avoid contamination when different chemicals are used. Tanks should have large filler necks to eliminate need for a funnel and facilitate cleaning. Use a removable fine-mesh screen to catch sediment.

Pumps — Centrifugal pumps handle all kinds of spray chemicals with least wear, but they push the liquid in one direction only. They must be mounted below the tank, or some priming arrangement must be provided. Turbine types develop higher pressures and move the spray in either direction, but they have to be mounted lower than the tank or be primed by hand. Gravity feeds are satisfactory for herbicides applied as coarse sprays or atomized by devices other than standard nozzles. The pressure and rate of flow decrease as the tank is emptied. Pump materials must be rust resistant. Operate the pump from wind-driven propellers, hydraulic motors, electric motors, or the accessory drive pad of the engine.

Pipes, fittings, and valves — Pipes are made of aluminum, stainless steel, or plastic. Hoses must be resistant to aromatic fuels, while fittings should be a dural or brass-flare type. Mainline valves should be quick closing; nozzles require nondribble shutoff valves.

Dusters and Spreaders

The use of dry materials, especially granular formulations instead of sprays, is increasing. Where water is not readily available, dusts are substituted for sprays and applied with various kinds of dusters from handoperated to power-driven machines. Dusters are simple to operate, can be filled rapidly, and travel at faster speeds than sprayers, but drift is a serious problem, and application of herbicidal dusts is prohibited by law in some states.

Granular products are more practical than dusts in most situations. Application is made by hand with broadcast seedsowers, fertilizer spreaders (both small and tractor-drawn), or with air-gun applicators. A typical hand spreader has a rotating plate powered by a handcrank for spreading the granules. It is carried by a shoulder sling strap, weighs 6 lb, has a capacity of 25 lb, and is adjustable for various rates. The rotary type of spreader handles heavy formulations effectively, but is not as satisfactory for formulations based on lighter materials such as vermiculite.

Granular herbicides, such as fenuron, are applied to brush by hand. Typical equipment consists of a 10-qt pail and plastic spoon for each member of a crew, and a 2-ton crawler tractor that pulls a scoot (a sled-type carrier with 6 in. steel runners for hauling bags of chemical). The tractor and scoot are driven along the highway, the crew (10 men) spread out across 70 ft, and as they walk, they throw a teaspoon of fenuron beside each brush cluster. Where brush is thick, the crew uses the grid system throwing a teaspoon of chemical every 3 ft.

REQUIREMENTS

Oils

All hose connections should be made of Teflon® or a neoprene material. The tank need not be heavy walled or equipped to operate under pressure. The pump should be milled with very close tolerances and should not require lubrication. Pumps that depend upon packing and repacking to prevent leakage are not satisfactory for application of oil.

Emulsions

The pump described for oils is satisfactory for emulsions. Internal rotary gear pumps, diaphragm, position, and nylon-roller types are adaptable to oil-water emulsions for spraying ditchbank weeds.

Suspensions

For small areas, use lawn sprinkling cans or knapsack sprayers and keep the spray mixture well agitated. With knapsack sprayers, use either cone-type or fan-type nozzles. Often small booms with two to four nozzles spaced 20 in. apart are useful. Skips or unnecessary overlapping can be avoided by marking off small areas to be covered at a time.

Use power equipment for large areas. Agitate the spray mixture continuously so that the chemical does not settle out. Mechanical agitation is usually adequate, but is ineffectual if the power unit is idled or operated at reduced speeds. Check the effectiveness of the agitation by watching for any powder left in the corners of the empty tank.

Effective suspension of water-dispersible powders can be obtained with hydraulic agitation provided from by-passed spray materials if the pump has sufficient capacity.

To construct a jet agitator, install a pipe on the nozzle supply side of the pressure regulator and extend it to and along the bottom of the tank to reduce foaming. One pint of kerosene per 100 gal spray is an effective antifoaming agent. Fit the pipe with sufficient jet-agitator nozzles or drill holes in the pipe to maintain turbulence. Recirculate the solution through this pipe at about 10% of the tank capacity per minute. A second return line, leading from the pressure regulator to the bottom of the tank, must also be used for pressure regulation. Use no finer than a 50-mesh screen or strainer when powders are used.

Gear pumps wear out quickly when used for suspensions of wettable powders. A piston pump, driven by a separate engine or diaphragm pump, is more dependable. Most portable units use a roller pump. Where tractors or trucks with power takeoff are to be used, a good roller or diaphragm pump is satisfactory.

Pumps for power-takeoff mounting deliver volumes roughly in proportion to the speed of the shaft. Standard maximum-power takeoff speed is about 540 r/min. If the speed of the power-takeoff shaft is not maintained near maximum, pump volume drops off. Since large pump capacities are necessary if the volume for both adequate spraying and agitation is to be maintained, it is important to spray with the vehicle throttle set as near maximum as possible. If the pump does not have enough capacity to supply both the agitator and the desired number of nozzles, reduce the ground speed of the vehicle by changing to a lower gear and use smaller nozzles demanding less volume.

Maintain a pressure range of 30 to 60 lb when a boom attachment is used. For handoperated spray guns with larger nozzles, increase the pressure. Since some small pumps have maximum outputs of 4 to 5 gal/min, it is impossible for them to recirculate 10% of a 55-gal tank per minute and furnish additional capacity for spraying. Use a pump with capacity sufficient to furnish 4 to 5 gal/min to the boom in addition to the volume needed for the agitator if hydraulic agitation is to be used.

A spray boom, mounted on a tractor or some other vehicle, provides one of the most accurate and rapid methods of application. A large area can be covered quickly, and the spray equipment can be calibrated accurately. In booms with large diameters, the spray liquid does not flow so rapidly as in small-diameter booms, and the powder may come out of suspension and settle in the boom. For average equipment, use booms 3/8 to 1 in. in diameter; the size depends on the length of the boom, the screen, and the size of the nozzles. For blanket spraying with nozzles at 20-in. spacing along the boom, the tips should be about 18 in. above the ground or tops of the weeds for 80° nozzles, and 21-in. lap of the spray pattern between nozzles, thus assuring a uniform and continuous swath.

Use a 50-mesh, or coarser, filter screen for the suction strainer and the line strainer. For the individual nozzles, use screens with only slightly smaller openings than the nozzle opening. Do not use flannel or cloth filters.

OPERATION

To a large extent, the proper adjustment and calibration of spray equipment determines the success of a spraying operation. Keep the nozzles clean and the volume per acre constant. Never use a metal object to clean nozzles; an old toothbrush cleans without injuring the orifice. The volume can be varied by changing the pressure, speed of travel, or nozzle size.

Regulating Volume

A change in pressure results in only small changes in volume. To double the volume, it is necessary to increase the pressure four times. The more convenient ways to change volume are to either change the speed of travel or the nozzle tip. Doubling the speed reduces volume one half. The amount of spray delivered by a nozzle is directly proportional to the square of the diameter of the opening. Thus, doubling the diameter of the opening increases the volume four times. Charts are available from spray equipment companies showing capacities of nozzle size in gallons per minute at different pressures; some charts give gallons per acre at designated speeds.

Even after selecting the proper nozzle, calibration is necessary. Nozzle output may vary considerably from the charted capacity. Calibration is not necessary where rates of application are given in pounds herbicide per 100 gal of spray, and the amount applied depends on density of foliage or stems, as in brush spraying with handguns.

Regulating Droplet Size

The smaller the droplets, the slower they fall, and the greater the chance of drift. The lower the pressure, the larger the average size of droplet. With a typical flat-fan nozzle, a decrease in pressure from 60 to 20 lb/in.^{2} can double the size of the average droplet. It can also decrease by four or five times the number of fine droplets.

To reduce the drift hazard, never use less than 10 gal/acre. Use 20- to 30-lb pressure for broad-leaved weeds. Keep nozzles as close as possible to the top of vegetation or the ground while maintaining a complete coverage within the swath. The boom may be lowered by rotating its axis so that nozzles spray backward or forward at an angle onto the foliage rather than straight down. A high ground speed allows use of larger nozzles for a given volume per acre. A high ground speed plus high gallonage plus low pressure permits the use of nozzles with large openings; hence, the droplets will be coarser and there will be less drift. Use a flooding type of nozzle if susceptible plants are nearby. Never increase output on the sprayer by raising the pressure; rather, change to a larger size nozzle. Use smoke trails to determine wind direction.

Calibration of Equipment

Both hand and power equipment require calibration for the conditions under which they are to be used. The manufacturer's calibration is made in the laboratory or factory and may not apply to field conditions. Furthermore, spray delivery changes as the machine becomes older — parts become worn and strainers and screens become partially clogged.

Hand Sprayer

The volume of spray per square rod or per 100 ft^2 in which the chemical can be applied depends on the size of the nozzle, nozzle spacing, and the speed at which the operator walks. To calibrate the sprayer, measure out a plot 1 rod^2 or 1000 ft^2 in area. Spray the plot with water in the equipment to be used and at the normal walking rate of the operator. Measure the water used. Convenient volumes for calculations are 1 pt/rod^2 (20 gal/acre) or 1 qt/rod^2 (40 gal/acre), or 2 qt/1000 ft^2 (22 gal/acre). The amount of chemical recommended for treatment is then mixed or dissolved in the volume of liquid thus determined.

Power Sprayers, Boom Type

Before proceeding to calibrate a sprayer, run water through the sprayer to see that all nozzles are clean and check to see if they discharge at a uniform rate by running water through them at a uniform pressure, catching the discharge from each nozzle in a separate container (such as a calibrated nursing or baby bottle). If the discharge varies widely, replace nozzle tips.

Determine the amount of herbicide to put in the tank by one of the following methods. The first involves determining the amount of liquid a sprayer applies per acre as follows:

1. Start with a tank filled to the brim with clean water. Adjust the pressure to what will be used in the field.
2. Drive exactly 1/8 mi (40 rods or 660 ft) in a field or along a road, ditchbank, or other area to be sprayed at the speed to be used when spraying — usually 3 to 5 mi/hr. Measure from where the spray begins, not where the tractor started. Operate the tractor at full throttle. Set speed by selecting right gear (so governor can maintain constant speed under variable loads).
3. Shut off the spray, return to the original filling position on level ground, and refill the tank. Measure the amount of liquid required.
4. Calculate the application rate as follows:

$$\frac{\text{Number of quarts used} \times 16.5}{\text{Width of spray swath (in feet)}} = \text{gallons per acre}$$

For example, if 6 qt of water were used in 1/8 mi and the spray width is 20 ft, multiply 6 by 16.5 and divide by 20. The result is 4.95, or about 5 gal/acre.

$$\frac{6 \times 16.5}{20} = \frac{99}{20} = 4.95 \text{ gal/acre}$$

Then divide the number of gallons the tank holds by the number of gallons the sprayer applies per acre. This gives the number of acres one filling will spray. Multiply the number of acres one tankful will spray by the amount of herbicide to be used for each tankful. For example, if the tank holds 55 gal and the sprayer applies 5 gal/acre, one tankful will spray 11 acres (55 divided by 5). If 1 pt of spray material is required per acre, 11 pt would be required for each tankful.

The second method is to catch the discharge from one nozzle in a pint jar as the sprayer is being operated at the speed and pressure that will be used for spraying. Measure the distance traveled (in feet) while collecting 1 pt. Then determine the rate of application per acre from Table 1. For example, if it takes 120 ft to collect 1 pt of spray and the nozzle spacing on the boom is 20 in., 27 gal/acre would be required.

A variation of Method 2 is to measure the time required to collect nozzle discharge. Partially fill the tank with water, turn on the sprayer, and set it to the desired pressure. Catch the total amount of spray discharged in a given number of minutes and divide the number of gallons by the number of minutes. This equals the output in gallons per minute. It may be more convenient to weigh than to measure the water collected; a gallon of water weighs 8.34 lb. Calculate the speed of travel as follows:

$$\frac{495 \times \text{gallons per minute}}{\text{Gallons per acre} \times \text{spray width (in feet)}} = \text{miles per hour}$$

For example, if the nozzle output is 2 gal/min, the volume of application desired is 50 gal/acre, and the spray width is 3 ft.

$$\frac{495 \times 2}{50 \times 3} = 6.6 \text{ or } 7 \text{ mi/hr}$$

If the calculated speed of travel is not practical, adjust the pressure, and calculate again.

Table 1
CALIBRATION OF SPRAYER BY COLLECTING THE DISCHARGE FROM ONE NOZZLE

Distance traveled to collect 1 pt (ft)	Gallons to be applied per acre when discharge equals 1 pt and nozzles are spaced at various intervals (in.)			
	20	18	15	12
40	82	91	109	136
50	65	73	87	110
60	54	60	73	91
70	47	52	62	80
80	41	45	55	68
90	36	40	49	62
100	33	36	44	55
110	30	33	40	50
120	27	30	37	46
130	25	28	34	42
140	23	26	31	39
150	22	24	29	36
160	20	22	28	34
180	18	20	24	31
200	17	18	22	28
220	15	16	20	25
240	14	15	18	23
260	13	14	17	21
280	12	13	16	20
300	11	12	15	18
400	5	6	7	9

Power Sprayers, Boomless Type

Calculated application data rather than calibration are used for boomless sprayers. These data are available from the technical data reports of nozzle manufacturers.

Aerial Sprayers

Determine the flow rate in gallons per minute in the following way. First, put a measured amount of spray in the tank or fill the tank to a definite level. Instruct the pilot to turn on the spray for a timed interval (30 to 60 sec) while flying level and straight at the speed to be used for spraying. When plane lands, drain and measure the liquid remaining in the tank or, with the plane in the location where the tank was filled, measure the amount required to refill the tank to the same level. Compute the flow rate in gallons per minute and the flow rate required as:

$$\frac{\text{SWD}}{\text{F} = 495}$$

where S = speed of plane in miles per hour; W = width of effective swath, not total swath in feet; D = dosage to be applied in gallons per acre; and F = flow rate in gallons per minute. For example, if S = 80 mi/hr, W = 40 ft, and D = 2 gal/acre:

$$\frac{80 \times 40 \times 2}{495} = 12.9 \text{ gal spray required per minute}$$

CLEANING AND PREPARATION FOR STORAGE

All spraying equipment should be cleaned after use. If the chemical is soluble in water, a thorough flushing and rinsing with water is sufficient. Equipment that has been used for phenoxy herbicides such as 2,4-D is very difficult to clean. Use one of the following methods for hand sprayers or field equipment:

1. Remove nozzles and scrub with kerosene
2. Ammonia rinse

    a. Add a box of nonsudsing detergent to 30 to 40 gal of water, run through the pump, by-pass for 5 min, and then run out through the boom
    b. Partly fill tank with a solution of 1 to 2% household ammonia (1 to 2 qt in 25 gal water, or 2 tsp/qt of water)
    c. Leave this solution in the sprayer (including hoses and boom) overnight
    d. Rinse thoroughly with clean water

3. Charcoal rinse

    a. Use at least one third of a tank of water. For each 10 gal of water add 1/4 lb of activated charcoal and 1/8 to 1/4 lb of laundry detergent. Agitate this mixture vigorously to distribute the charcoal through the water.
    b. Wash the equipment for 2 min by swirling the liquid around so that it reaches all parts of the tank. Pump some of the liquid through the hose and nozzles.
    c. Drain the tank and rinse the equipment with clean water.

Preparation for Storage

Scrub the sprayer with a stiff bristle brush. Coat all iron parts exposed to the chemical with a rust inhibitor or light oil. Remove nozzles, take apart, clean, and store in light oil. Before using the next year, wash in kerosene or gasoline to remove the oil.

1. Fill pump with a rust preventive.
2. Remove caps from the ends of booms and stand booms on end to remove sediment. Remove, clean, and reassemble filters.
3. If the sprayer is powered with a gas engine, drain fuel tank and carburetor and pour a tablespoon of engine oil through a spark plug hole. Turn motor over by hand to distribute oil on cylinder walls.
4. If the sprayer is to be stored outside, remove rubber hoses and keep in a cool dark place.
5. Thoroughly clean dusters and spreaders before storage.

REFERENCES

1. Department of the Army, Herbicide Manual for Noncropland Weeds, Technical Manual TM 5-629, U.S. Government Printing Office, Washington, D.C., 1970.
2. Department of the Navy, Basic Pest Control Technology, Naval Facilities Engineering Command, Alexandria, Va., 1976.
3. Department of the Navy, Chemical Control of Disease Vectors and Economic Pests, Naval Air Station, Jacksonville, Fla., 1976.
4. Stewart, L. E., Gienger, G. W., Parochetti, J. V., and Coan, R. M., Broadcast Spraying, Extension Bulletin 229, University of Maryland, College Park, 1976.

Chapter 10

HERBICIDE EFFICACY

INTRODUCTION

When using emulsions, the following inventory should be made: the location and quantity, growth conditions, width of right of way to be treated in various sections, location of access roads, ditch crossings, fence openings, accessible water for spraying foliage, and crops grown in the section that might limit the method or time of spraying. The salient considerations under job specifications are coverage, timing, and materials.

EMULSION SPRAYS FOR SUMMER APPLICATION

The Consumer Power Company, Jackson, Mich. has reported data on manual vs. chemical cost of vegetation control near power transmission lines from 1948 to 1965. Table 1 indicates that the amount of manual labor has been appreciably reduced over the years.

The ideal situation for the utilities' right-of-ways is a ground coverage of grasses or low-growing shrubs that would not interfere with maintenance. This condition is often difficult to obtain, especially where resistant species tend to crowd out the desirable cover. Once competition of the large growing woody plants has been removed, some sort of natural cover usually comes in. Therefore, the major problem is to kill the entire root system of the troublesome species. The results reported below are the summation of a 5-year study in which numerous chemicals and techniques of application were applied.

In 1955, several experiments were designed by Chappell[7] to study the effect of water sprays vs. applications of oil plus water. Treatments were made in May, June, August, and September using 6 lb of a 50 to 50 mixture of 2,4-dichlorophenoxy (2,4-D) and 2,4,5-trichlorophenoxy (2,4,5-T) per 100 gal of water. In another series, a mixture containing 10% Number 2 fuel oil was used. The results of this experiment are shown in Table 2. As may be seen from the data, the May and June treatments were much better than those of August and September. This and subsequent experiments show that the addition of oil to the spray solution was most effective after the middle of July. Treatments made earlier than this showed no response to the addition of 10% oil to the mixture.

Many of the so called "resistant species" are usually killed to ground level with little difficulty. Some of these species are sassafras, black locust, and sumac. It has been impossible to kill the root system of these species in one application of 2,4-D and 2,4,5-T during the growing season. Very often sprouting species will have an overall stem increase the second year after spraying. In 1957, following 2 years of study with various 2,4-D and 2,4,5-T esters, it became apparent that the most resistant species mentioned above could not be killed completely with these chemicals. During this time, screening tests showed that amitrole was very effective against these root-sprouting species. Comparison of amitrole at 8 lb/100 gal of water with 2,4-D and 2,4,5-T at 6 lb/100 gal of water are shown in Tables 3 to 6.

As indicated by the results, oak, maple, and other species were effectively controlled by the 2,4-D and 2,4,5-T esters and amines; however, black locust continually increased where this treatment was used. A further study using amitrole during May, June, and July showed that the July treatments were much more effective on all species than were the May or June treatments. Studies conducted in 1957 included monuron

Table 1
COST FIGURES FOR MAINTENANCE OF POWER TRANSMISSION AND SUBTRANSMISSION LINES — MANUAL AND CHEMICAL

Year	Total annual cost per mile ($)	
	Chemical	Manual and chemical
1948	3.13	27.33
1949	3.21	81.39
1950	11.13	73.31
1951	5.32	68.73
1952	13.80	55.87
1953	20.24	64.34
1954	22.68	50.08
1955	29.91	46.25
1956	46.95	58.19
1957	44.55	57.73
1958	16.02	26.44
1959	49.21	62.00
1960	50.17	66.27
1961	46.19	62.39
1962	38.34	52.76
1963	35.89	55.42
1964	47.66	63.96
1965	48.00	57.53

From Anon., Consumers Power Company, Jackson, Mich., 1965. With permission.

as a treatment for general brush control. These treatments were applied in June, August, October, and December in a granular form and in a slurry at 5 and 7½ g of active monuron per clump of brush. Although the data shows only 75 to 80% kill the first season, practically all the clumps of brush had died by the following year, 1958. Practically no root sprouting appeared on any of the species. While this chemical will kill almost any woody plant, it must be used with caution, since trees may be killed as far as 50 ft from where the application was made. Larger trees are much more susceptible than small plants; presumably, this is due to the much larger area of absorption in the trees.

As can be seen from the data in Table 5, black gum and sassafras were the two most resistant species in this experiment. Sumac and black locust were very susceptible. Of the treatments used, combinations of amitrole and 2,3,6-trichlorobenzoic acid (TBA) appeared most effective against root-sprouting species. The area chosen for these experiments was populated mainly with black locust, sassafras, and sumac with a few oaks and maples. These combinations of chemicals appear to be the most effective on these species to date (see Table 7).

Summer herbicidal applications are usually applied at pressures of 200 to 400 lb/in.^{2} at rates of 200 to 400 gal of solution per acre, depending upon the season of the brush being sprayed. Best results are obtained by the so-called "stem-foliage" application in which the entire stem and leaves are fully wetted. Complete coverage of the entire plants is necessary for good brush control.

Table 2
SUMMARY OF THE PERCENTAGES KILLED AND PERCENTAGES RESPROUTING[a] BY SPECIES FOR ALL TREATMENTS

Month of treatment	Red oak (%) Killed	Red oak (%) Resprout	Sassafras (%) Killed	Sassafras (%) Resprout	All species (%)[b] Killed	All species (%)[b] Resprout
May	86.2 a	22.3 a	100.0 a	61.9 a	94.1 a	44.1 a
June	59.8 b	61.0 b	100.0 a	46.6 a	87.8 b	42.3 a
August	10.0 c	10.2 c	92.3 a	126.9 a,b	77.5 c	46.5 a
September	2.0 c	3.8 c	72.0 b	172.6 b	63.7 c	59.6 a
Spray						
No oil	38.5 a	22.7 a	92.3 a	112.7 a	81.8 a	48.0 a
10% oil	36.2 a	22.7 a	87.5 a	110.0 a	79.9 a	48.0 a

Note: Two figures in a column followed by the same letter are not significantly different.

[a] Percentage resprouting was derived by dividing the number of resprouts by the number of sprouts before spraying and multiplying by 100.

[b] Except black locust (*Robinia pseudoacacia*).

From Chappell, W. E., Control of brush in powerline rights-of-way, paper presented to the National Meetings, American Society of Agricultural Engineers, Columbus, Ohio, June 1959. With permission.

Table 3
EFFECT OF TREATMENTS ON VARIOUS SPECIES — EXAMPLES OF INDIVIDUAL STEMS PER PLOT

Species	Amitrole (8 lb/acre) Before	Amitrole (8 lb/acre) After	2,4-D + 2,4,5-T ester (6 lb/acre) Before	2,4-D + 2,4,5-T ester (6 lb/acre) After	2,4-D + 2,4,5-T ester (6 lb/10 gal oil/90 gal water) Before	2,4-D + 2,4,5-T ester (6 lb/10 gal oil/90 gal water) After	2,4,5-T amine (6 lb/acre) Before	2,4,5-T amine (6 lb/acre) After
Red oak	18	4	107	55	131	66	535	91
Chestnut oak	31	6	49	29	119	85	99	19
Sassafras	28	144	36	59	51	261	95	12
Black locust	86	10	220	269	53	63	—	—
Sumac	—	—	—	—	39	0	—	—
Maple	—	—	113	56	52	35	—	—
Hickory	3	1	68	21	29	16	22	2
Black gum	3	49	—	—	4	1	40	0

From Chappell, W. E., Control of brush in powerline rights-of-way, paper presented to the National Meetings, American Society of Agricultural Engineers, Columbus, Ohio, June 1959. With permission.

Table 4
TIME OF AMITROLE APPLICATION

	Top kill (%)			Resprouts (%)		
Treatment	May	June	July	May	June	July
8 lb—100 gal	10	72.11	93.95	100	62.13	30.06
12 lb—100 gal	10	99.68	96.05	100	20.76	33.67

Note: In the area chosen for this test, the brush was 10 to 15 ft high and predominately black locust, sumac, sassafras, and oaks. The volume used was very high, but not uncommon on brush of this size. A breakdown of the major species in the plots is presented in Table 5.

From Chappell, W. E., Control of brush in powerline rights-of-way, paper presented to the National Meetings, American Society of Agricultural Engineers, Columbus, Ohio, June 1959. With permission.

Table 5
STEMS PER ACRE BEFORE AND AFTER TREATMENT

	8 lb Amitrole — June		12 lb Amitrole — June	
Species	Before	After	Before	After
Oak	49	15	93	2
Sassafras	28	144	198	127
Black gum	3	49	0	0
Black locust	86	10	254	3
Sumac	96	14	75	0

From Chappell, W. E., Control of brush in powerline rights-of-way, paper presented to the National Meetings, American Society of Agricultural Engineers, Columbus, Ohio, June 1959. With permission.

Table 6
APPLICATIONS OF MONURON

Treatment (g/clump)	Date	Reduction of stems (%)
Pellets (5)	6/4/57	79.80
Slurry (5)	8/5/57	79.25
Slurry (7.5)	8/5/57	74.86
Pellets (5)	10/1/57	74.06
Pellets (7.5)	10/1/57	74.65
Pellets (5)	12/19/57	72.51
Pellets (7.5)	12/19/57	73.63

From Chappell, W. E., Control of brush in powerline rights-of-way, paper presented to the National Meetings, American Federation of Agricultural Engineers, Columbus, Ohio, June 1959. With permission.

Table 7
TREATMENTS AND RESULTS OF EXPERIMENTS ON BLACK LOCUST

Treatments (per 100 gal of water)	Resprouts (%)	Original stems alive (%)
Amine D and T (4 lb/acre)	60	6
Amine D and T 4 + ATA (1 lb/acre) (ACP 329)	21	0
2,3,6-TBA amine (duPont) (8 lb/acre)	15	1
2,3,6-TBA amine (duPont) 2 + ATA (2 lb/acre)	20	0
2,3,6-TBA amine (duPont) 4 + ATA (2 lb/acre)	2	2
2,3,6-TBA amine (duPont) 4 + ATA (4 lb/acre)	5	5
2,3,6-TBA amine (duPont) 8 + ATA (2 lb/acre)	0	1
2,3,6-TBA amine (duPont) 8 ATA (4 lb/acre)	10	1
2,3,6-TBA amine (duPont) 8 + ATA (8 lb/acre)	6	0
2,3,6-TBA esters (ACP) 4 + ATA (1 lb/acre)	4	0
2,3,6-TBA esters (ACP) (8 lb/acre)	7	0
ATA (8 lb/acre)	6	0
Butoxy ethanol esters of D and T (4 lb/acre)	35	4
Butoxy ethanol esters of D and T (4 lb/10 gal oil/90 gal water)	36	4
Ethyl hexyl esters of D and T (4 lb/acre)	25	2

From Chappell, W. E., Control of brush in powerline rights-of-way, paper presented to the National Meetings, American Society of Agricultural Engineers, Columbus, Ohio, June 1959. With permission.

Table 8
CANE BROADCAST TREATMENTS
SPRAYED OCTOBER 1956

Treatments (per 10 gal oil)	lb/acre	Stems dead to ground		Resprouts	
		No.	%	No.	%
Tech D and T	8	12,452	97.4	335	2.6
Tech D and T	4	13,712	94.8	754	5.2
Tech T	8	7,695	92.3	632	8.5
Tech T	4	10,709	97.9	230	2.1
Tech T	2	10,627	78.2	2,961	21.8
IBK	8	10,011	93.9	65	6.1
IBK	4	5,207	69.3	2,310	30.7
236 TBA	9	7,887	76.6	2,413	23.4
Tech D and T	4	13,712	94.8	754	5.2
Tech T	4	10,709	97.9	230	2.1

From Chappell, W. E., Control of brush in powerline rights-of-way, paper presented to the National Meetings, American Society of Agricultural Engineers, Columbus, Ohio, June 1959. With permission.

DORMANT OR WINTER SPRAY APPLICATIONS

The practice of basal spraying with 2,4,5-T in oil during the winter months has been used for some time. However, this practice has not been adopted in general brush control near power lines. It is a time-consuming method, and often smaller sprouts on lines are missed by this application method. While attempting to apply some mixtures by this method in October 1956 to a right-of-way heavily invested with many species

Table 9
FALL APPLICATIONS USING THE BROADCAST BASAL TECHNIQUE, OCTOBER 10, 1958

Treatments	Resprouts[a] (%)	Original stems[a] alive (%)
Tech T (6 lb/acre), oil cane broadcast and soil[b]	24	0
Tech T (6 lb/acre), oil broadcast basal and soil[c]	8	1
Weedone® T (6 lb/acre), oil cane broadcast and soil	20	0
Weedone® T (6 lb/acre) oil broadcast basal and soil	24	0
Industrial BK 8-2080, oil cane broadcast and soil	37	1
Industrial BK 8-2080, broadcast basal and soil	41	1
Check (untreated)	0	100

[a] Average of two replicates.

[b] Cane broadcast and soil — The stems wet as usual, and the soil sprayed around root-sprouting species such as locust, sassafras, and sumac.

[c] Broadcast basal and soil — The lower 3 ft of stems sprayed thoroughly, and the soil sprayed around root-sprouting species.

From Chappell, W. E., Control of brush in powerline rights-of-way, paper presented to the National Meetings, American Society of Agricultural Engineers, Columbus, Ohio, June 1959. With permission.

of woody plants, lower concentrations were effective and cheaper to apply, even though considerably more volume was necessary. It was further noted that no emulsifier was necessary to mix 100% of technical esters with oil and that a technical ester should be suitable for this oil application.

In another study, technical esters of 2,4,5-T were applied at 2, 4, and 8 lb of acid per 100 gal of Number 2 fuel oil. Application volumes in this test were approximately 125 gal/acre. The initial results of the experiment are shown in Table 8. Various esters of 2,4-D and 2,4,5-T were compared in 1957 and 1958, applied by different techniques at different times during the dormant season from September to April. The September treatments were ineffective; it can be assumed that much of the chemical was absorbed by the leaves where they were burned off by the oil, and no translocation occurred. On October 10 at elevations around 2000 ft in western Virginia, most of the leaves are off of deciduous plants, leaving the stems exposed. At that time, they are easily covered with the oil mixture. These experiments showed that by applying most of the chemicals around the base of plants, and even on the soil near root-sprouting species during cold weather, a higher percentage kill can be obtained than by the cane broadcast method. During the winter of 1958 to 1959, several chemicals were compared using what is termed the broadcast basal technique. Tables 9 and 10 show the results from the fall and spring applications using this method.

Dormant applications have many obvious advantages in spraying right of ways.

1. There is very little danger of injury to crops in the October to April season.
2. About one half of the volume is necessary for good coverage as compared to summer treatments.
3. Lower pressure and, consequently, lighter hoses and smaller crews are needed for this application.
4. It is much easier to get through and apply the chemicals when no leaves are on the brush.
5. There is much less brownout than occurs with summer applications.

Table 10
SPRING APPLICATIONS USING THE BROADCAST BASAL TECHNIQUE, APRIL 1959

Treatments	Resprouts[a] (%)	Original stems[a] alive (%)
Tech T (6 lb/acre), oil broadcast basal	4	1
Tech T (9 lb/acre), oil broadcast basal	6	1
Tech T (12 lb/acre), broadcast basal	6	2
Diamond N-071 (6 lb/acre), braodcast basal	6	2
Trinoxol® (6 lb/acre), broadcast basal	11	1
Trinoxol® (9 lb/acre), broadcast basal	17	1
Trinoxol® (12 lb/acre), broadcast basal	4	1
Check (untreated)	0	100

[a] Average of two replications.

From Chappell, W. E., Control of brush in powerline rights-of-way, paper presented to the National Meeting, American Society of Agricultural Engineers, Columbus, Ohio, June 1959. With permission.

One obvious disadvantage is the physical contact with the oil on the applicator which may irritate the skin. This can be largely overcome by the use of light-weight neoprene suits.

Of the materials used in the dormant cane broadcast applications, 2,4,5-T esters and oil-soluble amines have shown the most promise. In general, formulations containing 6 lb of water per gallon are suggested because less volume of the concentrate is required.

SPECIFIC APPLICATIONS

The varying soil types, weather conditions, and brush species prevalent over the country make it impossible to set up a program of brush spraying that would be applicable everywhere. However, most of the chemicals mentioned in this chapter have been used successfully over most of the northeast region. Presently it appears that the following schedule is the most effective known procedure.

Initially cleared rights-of-way — Allow vegetation to grow for one full season and apply 2,4,5-T ester or an oil-soluble amine at a rate of 6 lb acid per 100 gal of Number 2 fuel oil as a dormant broadcast basal and soil application during the period November 1 to April 1 in the northeastern region.

Rights-of-way that have been cleared 2 to 3 years and have not been sprayed — Use a stem-foliage application any time after full-leaf development until about August 1. Usually, the addition of 10% fuel oil is beneficial after July 15. A concentration of 4 lb/100 gal of water of 2,4-D and 2,4,5-T low-volatile ester should be used for this application.

Lines that have been sprayed several times that are predominately covered with sassafras, black locust, sumac, and ash — Use a summer stem-foliage application of 4 lb of amitrole plus 4 lb of TBA per 100 gal of water. This should be applied any time after leaves are fully developed until about August 1.

REFERENCES

1. **Grundy, W. M. and Bennett, J. M.,** Aerial application of herbicides for right-of-way brush control, *Down Earth,* 16(2), 6, 1960.
2. **Mann, R. A.,** Stump treatment after initial clearing on right-of-way of newly constructed lines, *Proc. South. Weed Conf.,* 16, 318, 1963.
3. **Mann, R. A.,** Pellet application by helicopter on TVA rights-of-way, *Proc. South. Weed Conf.,* 19, 307, 1966.
4. **Mann, R. A. and Aldred, J. R.,** Brush control of utility rights-of-way by dormant application, *Proc. South. Weed Conf.,* 17, 295, 1964.
5. **Nation, H. A.,** Woody plant control on utility rights-of-way with "Tordon" herbicide pellets, *Proc. South Weed Conf.,* 18, 387, 1965.
6. **Coffey, W. J.,** Planning and programming and brush control, *Down Earth,* 11(1), 10, 1955.
7. **Chappell, W. E.,** Control of brush in powerline rights-of-way, paper presented to the National Meetings, American Society of Agricultural Engineers, Columbus, Ohio, June 1959.
8. **Bovey, R. W.,** Response of Selected Woody Plants in the United States to Herbicides, Handbook NO. 493, U.S. Government Printing Office, Washington, D.C., 1975.
9. **Anon.,** Consumers Power Company, Jackson, Mich., 1965.

Chapter 11

INDIVIDUAL STEM TREATMENT

INTRODUCTION

Picloram (4-amino-3,5,6-trichloropicolinic acid) is a new herbicide known by the trademark Tordon®. It has shown wide spread promise for the control of many woody plant species, as well as herbaceous perennial weeds by conventional foliage sprays.[1-6] Field studies were conducted over a 2-year period to evaluate Tordon® for use in basal frills or in tree injectors for the control of certain hardwood species. Tordon® formulated as the potassium or triisopropyl amine salt was used in these studies.[7]

BASAL FRILL STUDIES

Single-stem white oak trees (*Quercus alba* L.), ranging in size from 5 to 10 in. in diameter at 5-ft height, were treated in April. They were still dormant and had complete and half frills located both 2 to 4 in. and 3 ft above the ground. With each type of frill treatment, the cut surface was thoroughly wet with the herbicide solution. The treatments were applied with a small, compressed-air sprayer equipped with a narrow angle cone nozzle. Chemical treatments were applied immediately after cutting the frills with a hand axe. The cuts were overlapped to give a continuous frill, and the depth of cutting was approximately 1/2 in. Five randomly selected trees were used with each treatment.

Results were evaluated near the end of the second growing season after treatment. Each tree was treated using the following numerical rating system: 1 = little or no effect; 2 = some effect, but less than 50% top kill; 3 = over 50%, but not complete top kill; 4 = complete top kill, but resprouting at or near root collar; and 5 = complete top kill with no resprouting. Results are expressed as percent of treated trees falling in each control class.

Complete Basal Frill

Tordon® herbicide diluted in water was extremely effective when applied in a complete continuous cut frill located 2 to 4 in. above the ground. The results of this method of treating, as evaluated 16 months after application, are presented in Table 1. Tordon® killed all trees at all concentrations, while two out of five trees resprouted from the crown with each of the standard 2,4,5-trichlorophenoxy (2,4,5-T) treatments.

Complete Frill 3 ft Above Ground

With the complete frill 3 ft high, results with Tordon® were similar to those obtained with basal frills. However, 2,4,5-T ester in fuel oil was much less effective than when used in a frill at the base of the tree. The data in Table 2 indicate that, at 2.9-lb acid equivalent (ae) per 100 gal in water, Tordon® gave complete kill on all trees with no resprouting. However, 2,4,5-T in oil at 8-lb acid equivalent per 100 gal in oil killed those portions of the trees above the frill, but did not adequately translocate downward to kill the lower stems and roots.

Continuous Frill

Treatments applied to frills made half way around the tree and on one side only were relatively ineffective, regardless of chemical or frill position. Data are presented

Table 1
RESPONSE OF WHITE OAK TREES TO AQUEOUS SOLUTIONS OF TORDON® APPLIED TO A COMPLETE BASAL FRILL

Treatment	Active ingredient (lb/100 gal)	Trees in each control class (%) Diameter (in.) 1	2	3	4	5
Tordon® in water	0.7	—	—	—	—	100
Tordon® in water	1.4	—	—	—	—	100
Tordon® in water	2.9	—	—	—	—	100
Tordon® in water	5.8	—	—	—	—	100
1,4,5-T in fuel oil	4.0	—	—	—	40	60
2,4,5-T in fuel oil	8.0	—	—	—	40	60
Fuel oil only	—	—	40	—	60	—

From Watson, A. J. and Mesler, R. J., Jr., *Down Earth*, 19, 20, 1964. With permission.

Table 2
RESPONSE OF WHITE OAK TREES TO TORDON® AND 2,4,5-T APPLIED TO A COMPLETE FRILL 3 FT ABOVE THE GROUND

Treatment	Active ingredient (lb/100 gal)	Trees in each control class (%) Diameter (in.) 1	2	3	4	5
Tordon® in water	2.9	—	—	—	—	100
2,4,5-T in fuel oil	8.0	—	—	—	100	—

From Watson, A. J. and Mesler, R. J., Jr., *Down Earth*, 19, 20, 1964. With permission.

in Table 3. Although the difference between chemicals was much less marked than with complete frills, Tordon® was somewhat more effective than 2,4,5-T.

TREE INJECTION STUDIES

Two exploratory field experiments[7] were established in April to evaluate Tordon® when injected into certain hardwood trees. A commercial tree injector with a 1¼ in-wide cutting blade was used. Factors considered were the dosage per injection and spacing of injections. At time of treating, the trees appeared to be dormant. Leaf buds

Table 3
RESPONSE OF WHITE OAK TREES TO TORDON® AND 2,4,5-T PLACED IN A FRILL CUT HALF WAY AROUND THE TREE

Treatment	Active ingredient (lb/100 gal)	Location of frill	Trees in each control class (%) Diameter (in.) 1	2	3	4	5
Tordon® in water	2.9	Base	—	40	40	—	20
2,4,5-T in fuel oil	8.0	Base	100	—	—	—	—
Tordon® in water	2.9	3 ft	—	100	—	—	—
2,4,5-T in fuel oil	8.0	3 ft	80	20	—	—	—

From Watson, A. J. and Mesler, R. J., Jr., *Down Earth,* 19, 20, 1964. With permission.

had not started to open, but sap flow was under way as evidenced by profuse bleeding of the maple trees following back injury.

Dosage Experiment

An attempt was made to determine the minimum amount of Tordon® required per injection with injections spaced on 3-in. centers around the base of hardwood trees. Since the cutting edge of the injector blade was 1¼ in. wide, an uncut strip about 1¾ in. wide remained between injection points. Five single-stem trees were used per treatment: two American elm (*Ulmus americana* L.), one or two maple (*Acer saccharum* Marsh.), and one or two swamp white oak (*Q. bicolor* Wild). The trees ranges from 5 to 12 in. in diameter at 5-ft height.

A formulation of Tordon® containing 1.4 lb active chemical per gallon was used undiluted in the injector where a 1.4 lb/gal formulation is indicated. This formulation was also diluted 1:1 with water and used as a 0.7-lb/gal formulation. A 4-lb/gal triethylamine salt formation of 2,4,5-T was included as a commercial standard.

The results at the end of the first growing season were as indicated in Table 4. American elm and swamp white oak were controlled very effectively at all concentrations; amounts as low as 0.25 mℓ of a 0.7-lb/gal formulation (21 mℓ of Tordon® per injection) gave a complete top-kill with no resprouting. In contrast, 120 to 240 mg of 2,4,5-T were required for equivalent control. Maple trees were not controlled by any treatment due presumably to washing the herbicide from the injection points by profuse sap flow.

Spacing of Injections

In another experiment, a group of white ash trees (*Fraxinus americana* L.), ranging from 4 to 7 in. in diameter at 5-ft height, were selected to determine optimum spacing of injections; compare 4-ft-high placement injections with those placed near the ground line. Treatments were applied in early April with a commercial tree injector having a cutting blade 1¼ in. wide. A potassium salt formulation containing 1.4 lb. of Tordon® per gallon was used. The height and spacing of injections and results at the end of the first growing season are shown in Table 5.

Table 4
RESPONSE OF AMERICAN ELM, SWAMP WHITE OAK, AND RED MAPLE TREES TO INJECTIONS OF TORDON® AND 2,4,5-T

Herbicide	Conc. (lb ae/gal)	ml/injection	Toxicant/injection (mg)	Species	1	2	3	3	5
Tordon®	0.7	0.25	21	Elm	—	—	—	—	100
				Oak	—	—	—	—	100
				Maple	100	—	—	—	—
		0.50	42	Elm	—	—	—	—	100
				Oak	—	—	—	—	100
				Maple	100	—	—	—	—
		1.00	84	Elm	—	—	—	—	100
				Oak	—	—	—	—	100
				Maple	50	—	50	—	—
	1.4	0.25	42	Elm	—	—	—	—	100
				Oak	—	—	—	50	50
				Maple	100	—	—	—	—
		0.50	84	Elm	—	—	—	—	100
				Oak	—	—	—	—	100
				Maple	100	—	—	—	—
		1.00	168	Elm	—	—	—	—	100
				Oak	—	—	—	—	100
				Maple	100	—	—	—	—
2,4,5-T	4.0	0.25	120	Elm	100	—	—	—	—
				Oak	—	100	—	—	—
				Maple	100	—	—	—	—
		0.50	240	Elm	—	—	—	—	100
				Oak	—	—	—	—	100
				Maple	100	—	—	—	—

From Watson, A. J. and Mesler, R. J., Jr., *Down Earth,* 19, 20, 1964. With permission.

Table 5
RESPONSE OF WHITE ASH TREES TO INJECTIONS OF TORDON®

Height and spacing of injections	ml/injection (¼ lb/gal formulation)	mg/injection	Trees in each control class (%) 1	2	3	4	5
Injections made at base of tree							
Complete frill	1.0	168	—	—	—	—	100
	0.5	84	—	—	—	—	100
Basal (2 in. apart)	0.5	84	—	—	—	—	100
Basal (4 in. apart)	0.5	84	—	—	—	20	80
Basal (6 in. apart)	0.5	84	—	—	—	100	—
Basal (7 in. apart)	0.5	84	—	—	100	—	—
Quartered (4 in. equally spaced)	1.0	168	—	—	—	100	—
Quartered (4 in. equally spaced)	0.5	84	—	—	50	—	50
Halved (2 in. equally spaced)	0.5	84	—	50	50	—	—
Injections made 4 ft high,							
Complete frill	0.5	84	—	—	—	100	—
Quartered (4 in. equally spaced)	0.5	84	—	—	—	100	—
Halved (2 in., equally spaced)	0.5	84	—	50	50	—	—

From Watson, A. J. and Mesler, R. J., Jr. *Down Earth,* 19, 20, 1964. With permission.

The current results on white ash suggest the following conclusions:

1. Injections made near the base of a tree are more effective than those placed 4 ft high.
2. Maximum effective spacing of injections, based on a dosage of 84 mg per injection (0.5 mℓ of 1.4-lb/gal formulation), appears to be between 2 and 4 in. center to center. It is recognized that optimum spacing probably will vary with species and, to some extent, with the dosage per injection.
3. Equally spaced injections (2- to 4-in.) on white ash trees, 4 to 7 in. diameter, up to 84 mg per injection, regardless of height of injections, appear to be most effective.
4. Injections 4 ft high may give complete top-kill, but white ash trees tend to resprout below the treated area.

SUMMARY AND CONCLUSIONS

These studies indicate that Tordon® herbicide is considerably more effective than 2,4,5-T when applied to cut surfaces of hardwood trees when either sprayed into a cut frill or applied with a tree injector. Tordon® is readily translocated both upward and downward from the point of injection. It appears to have a distinct advantage over 2,4,5-T in its downward translocation into the root systems and in preventing basal resprouting.

Limited lateral translocation occurs, but treatments applied to only one side of a tree were relatively ineffective. On white ash, the maximum effective spacing with a tree injector, using a dosage of 84 mg per injection, appears to be between 2 and 4 in. on center. Injection treatments of maple during spring sap flow were not effective.

REFERENCES

1. **Klingman, G. C. and Ashton, F. M.,** *Weed Science; Principles and Practices,* Interscience, New York, 1975.
2. Department of the Army, Herbicide Manual for Noncropland Weeds, U.S. Government Printing Office, Washington, D.C., 1970.
3. Department of the Navy, Basic Pest Control Technology, Naval Facilities Engineering Command, Alexandria, Va., 1976.
4. *Farm Chemicals, Pesticide Dictionary,* Meister, Willoughby, Ohio, 1975.
5. **Bovey, R. W.,** Response of Selected Wood Plants in the United States to Herbicides, Agriculture Handbook No. 493, U.S. Government Printing Office, Washington, D.C., 1975.
6. **Bovey, R. W. and McCarthy, M. K.,** Establishment of fire-breaks on forest and rangeland with herbicides, *J. Range Manage.,* 18 (5), 282, 1965.
7. **Watson, A. J. and Mesler, R. J., Jr.,** Effects of Tordon® herbicide as basal frill and tree injection treatments in certain hardwood trees, *Down Earth,* 19, 20, 1964.

Part III: Research Studies Of Herbicide Applications

Chapter 12

PROGRAMS OF RIGHT-OF-WAY CONTROL

INTRODUCTION

In addition to maintaining the roadway and shoulders, highway maintenance must face the problem of vegetation control on the roadside. For many years, this task was handled solely by mechnical means. The use of herbicides on roadways came into practice about 1950, the same time that they were put to use on utility railroad, pipeline, and firebreak rights-of-ways. From the standpoint of maintenance, spraying is definitely justified.[1] Once the weeds and brush are eliminated, the prevailing grasses are given a chance to establish themselves. For roadsides with banks that are steeply sloped, full of rocks and debris, and to which one cannot gain access with mowing equipment, chemical spraying of herbicides is the only practicable method of weed control.

CONVENTIONAL METHODS OF CONTROL

Highways

Zukel and Eddy[2] have discussed the use of herbicides on highways, and the following reasons are given for their use:

1. To increase highway safety by eliminating tall weeds; thus reducing driving hazards by allowing an unobstructed view of warning lights, traffic lights, crossroads, driveways, and the view of curves
2. To control broad-leaf weeds, with the primary purpose of reducing the number of mowings
3. To keep the water drainage areas free of weeds and brush which impede water flow
4. To reduce the weed population as a source of infestation to neighboring crops
5. To eliminate toxic plants such as poison ivy
6. To eliminate ragweed and other allergy-producing weeds from the right of way

Zukel and Eddy point out of some of the hazards associated with herbicide applications along roadways.

1. Drift
2. Volatility
3. Washoff of insoluble soil sterilants
4. Leaching of soil sterilants
5. Public reaction to high degrees of brownout

The utility services, such as telephone and power companies, have an interest in the control of roadside brush. Unwanted vegetation can seriously interfere with power and telephone services by shorting lines and must be controlled. Many different types of herbicides are used for vegetation control along roadways.

Railroads

In many respects, effective control of vegetation on railroad rights-of-way is similar to that of roadsides. More soil sterilants come into use because the railroad ballast is also treated. There are two general problem areas of weed control along railroad rights-of-way: (1) the track or roadbed and (2) the off-track area.

Parker[3] and Yazell[4] give the rollowing reasons for vegetation control:

1. In ballast section; to keep the roadbed drained properly, enable ties and fastenings to be inspected, increase the life of the tie, and to keep weeds from traction motors and rails to prevent wheel slippage
2. On the shoulders, to maintain proper drainage and unrestricted sight, thus enabling inspection of trains in operation
3. Around bridges and structures for fire prevention
4. In the yards for convenience and appearance
5. On noxious weeds as covered by state laws
6. Around low switch stands and dwarf signals for safety
7. Around the inside of curves to maintain line of sight
8. Under communication lines for uninterrupted service
9. Adjacent to tracks to keep from fouling trains
10. At highway grade crossings to improve sight for both autos and trains
11. Around signs (mile posts, whistle boards) for line of sight

Historically, weed and grass control methods along railroad rights-of-way parallel that of roadways and include manual removal, burning, rotary mowing, and chemical control. Burning is not selective and is almost as expensive as chemical control. Furthermore, it does not destroy the root systems and creates a fire hazard. This method is now considered undesirable and is rapidly being replaced by other methods. Rotary mowing is still extensively practiced at crossings and along rights-of-way because local ordinances frequently prescribe it. However, as local ordinances are changed and better techniques are available, many railroads are replacing mowing with chemical treatment.

Along railroad rights-of-way, chemical control ranges from bare ground to selective weeding. Complete control on bare ground is expensive and requires high rates of soil sterilizing chemicals followed by reduced rates of materials with long residues. Since several railroads span most of the U.S., they have special problems for vegetation control of a widely varying nature. Yazell[4] has pointed out that continued usage of single chemicals will result in the encroachment of other types of vegetation which have to be dealt with by different herbicides. Examples are

1. The use of chlorates in the southwest controls Johnsongrass, but bermudagrass and vines flourish.
2. Continued use in the midwest of contact materials plus residuals at low dosage has resulted in the encroachment of bermudagrass, bindweed, giant smart weeds, horse tail, and chord grass.
3. In areas of low rainfall (15 to 20 in.), annuals are replaced by the more resistant bindweed, Indian grass, buffalo grass, bluestem, and Johnsongrass.

Control of woody plants along railroad rights-of-way is very important and in the past has been a rather difficult and expensive part of a maintenance program. Dormant treatment of brush with brush killers has been more popular in some areas since it offers a high percentage of control. The application can be made in the time of the year when the susceptible crop is not growing. Brush-control methods currently used on railroads include dormant sprays, basal application, modified basal application, cut stump, frill, and girdle-tree injections.

An example of a weed-control program for railroads that would be representative for the southern part of the U.S. has been reported by Lowry,[5] who was concerned with practices on the Cotton Belt Railroad right-of-way. They sprayed 2,4-D and 2,4,5-T on noncrop areas and ammate to adjacent crop lands in late summer and found that good coverage of foliage is absolutely necessary. Usually, no better than an 85% kill is expected. The resistant species on their trackline are sumac, persimmon, elderberry, green ash, red maple, and buttonwood.

Francisco and Aldred[6] have discussed the program used by the Tennessee Valley Authority (TVA) in controlling vegetation around their steam plants. The plants have an average of 25 mi of railroad which consist of line, switchyards, transformer yards, and approximately 20 aces inside the plant (fences, riprap on intake, and discharge tunnels, etc.). It is difficult to select chemicals for vegetation control in these areas because of two characteristics of the vegetation: (1) deep penetrating root systems and (2) the plants' ability to spread. Examples of herbicides and rates of application that have been tested are as follows:

1. Railroads
   a. Borate chlorate mixture (35% active chlorate), 6 lb/100 ft^2, for control of all vegetation
   b. Ammate, 300 to 350 lb/acre, for kudzu control
   c. Trysben, 4 gal in 100 gal of water, for control of conifers, persimmon, sassafras, black locust, and willows
   d. 2,4-D, 6 lb/acre, for control of trumpet vine
2. Switchyards, transformer yards, and fences
   a. Telvar-W, 50 lb acid equivalent (ae) per acre, for control of all vegetation
   b. Sodium TCA, 100 lb/acre, for control of Johnsongrass

As a result of extensive field testing by TVA, the following conclusions have been drawn:

1. Because of the many types of vegetation established, a combination of chemicals is required.
2. The wet sprays are more expensive and less effective.
3. Early applications are the most effective.
4. Chemical mixtures, either pellets or granular, applied by mechanical spreaders are the most economical.

TVA experience indicates that after the first year of treatment, only spot treatment is needed in the second and third year. They found that the most troublesome vegetation to recur was trumpet creeper, honeysuckle, briars, and crabgrass. The best combination for spot treatment was a mixture containing 80% disodium tetraborate pentahydrate, 13% disodium tetraborate decahydrate (boron trioxide equivalent, 43%), 3% 3 (*p*-chlorphenyl)-1,1-dimethylurea (monuron), and 1% trichlorobenzoic acid (TBA).

Pipelines

One of the reasons for the small amount of brush control on pipeline rights-of-way has been the history of poor public relations between companies and land owners.[7] In 1940, the Justice Department brought suit against virtually all the pipeline companies in the U.S. charging excessively high tariffs which resulted in refunds and rebates to shipper-owners under the guise of earnings or dividends. Improved technology, expanding petroleum demands, and in general, the realization of their public relations image caused most companies to alter their negative public relations attitude and provide proper weed control on pipeline rights-of-way.

The petroleum industry has a need for pipeline vegetation control to:

1. Remove fire hazards
2. Provide aerial surveillance of pipeline for detection of breaks
3. Control noxious weeds
4. Establish good public relations

Control is needed around the following areas connected with the recovery and transport of crude oil:

1. Christmas tree valves and piping at the well site
2. Tank battery, a group of storage tanks that the crude oil flows into from the well head or Christmas tree
3. Related facilities
   a. Heaters
   b. Valve mainfolds
      1. Compressors
      2. Field offices
      3. Gas plants

The sequence of a mangement program for the control of vegetation in these areas would include:

1. A survey to determine soil types, vegetation density, and type
2. The formulation of a program to use the proper types of herbicides
3. Trained personnel for application
4. Adequate records so that one can evaluate the cost of controlling the weeds.

Prior to World War II, vegetation was controlled mechanically, using crews of men with hoes, rakes, and cutting devices to remove the weeds on all company property. Other methods were tried such as the use of oil and salt water. Crude oil was not satisfactory because the concentrations required to sterilize soil conditions constituted a fire hazard. As a result of the Humble Oil Company's experience with soil sterilants,[8] they have settled on an initial treatment consisting of 30 lb/acre of diuron or monuron mixed with a surfactant and 400 gal of water per acre. This application was followed with spot retreatment in July which consisted of 30 lb/acre of bromacil with surfactant and TCA mixed in 400 gal of water. Some of the cost evaluations are

1. Hand control of weeds, depending upon site and location, costs from $300 to $800/acre
2. The use of soil sterilants, costs from $100 to $275/acre
3. The cost of application of monuron or diuron in the above described procedure, including factors that influence the treatment cost are type of soil, type of density in growth, and density in number of facilities treated (The total treatment costs varied from $51 to $192/acre, with labor costing $25 to $88/acre and material from $25 to $148/acre)

Firebreaks

Firebreaks are a convenient method to limit fire spread in forests and on range land.[9] This is usually done by cultivating a soil barrier around the designated area. There are

FIGURE 1. Selective vegetation management at Waveland Park entrance by mowing and restricted use of soil sterilant to control unwanted vegetation. (Courtesy of Little Rock District, U.S. Army Corps of Engineers, Arkansas.)

some inherent disadvantages in this technique in that it exposes the soil to wind and water erosion. The common soil sterilants such as monuron, diuron, borates, and bromocil are relatively effective, even in the dry climates, but high rates are required. They will persist up to 10 years under low rainfall, degrading about ½ lb/acre/month.

As a basic part of many rights-of-way vegetation control programs, more extensive use of dormant sprays of the phenoxy and picloram herbicides is being made. This has the advantage of using highly technical personnel in the low-peak labor season and reducing the image of brownout that follows treatment when the shrubs and trees die. It also has the advantage of application during a season of the year when there is little possible damage to desirable plants such as crops and vegetables.[10]

A relatively new herbicide, glyphosate (*N*-phosphonomethyl glycine), has been developed which is very effective for control of grassy perennials and small shrubs.[11] It is a broad-spectrum herbicide registered for industrial uses, including railroad rights-of-way, highways and roadsides, and fuel and power transmission lines. The degree of control depends to a large extent on the time and rate of application. Most common weeds are controlled with 1 to 3 k/ha.

SELECTIVE VEGETATION MANAGEMENT AND MANIPULATION

Mechanical Control

Management of rights-of-way vegetation is most effectively done by a combination of ecological and agronomic principles. The vegetation, potential or existing, should be predominantly grass, grass-like plants, legumes, forbs, or low shrubs. These are the climax vegetation of natural grasslands, shrublands, and alpine communities or coastal marshes, including nonproductive areas such as desert or even barren rock.[11]

In the development of new transport facilities, it is often possible to design the right-of-way so that vegetation control is more manageable, i.e., slopes are properly selected and graded to minimize cut and fill, and natural stabilized channels are used for waterways and drainage to avoid severe erosion potential. By limiting stripping, grubbing, and clearing operations during construction, more indigenous vegetation is conserved, and existing trees, shrubs, and turf are left undisturbed (see Figures 1 and 2).

FIGURE 2. Selective vegetation management at Waveland Park sanitary dump station by mowing and careful application of soil sterlant around the cleanup areas. (Courtesy of the Little Rock District, U.S. Army Corps of Engineers, Arkansas.)

Wherever the soil and ground cover are disturbed, it is important to till, fertilize, seed, and mulch the soil as soon as possible to prevent erosion. Special plant species, such as perennial legumes and adapted grasses, should be used for revegetation. Seeding mixtures including a temporary annual may be necessary to accommodate specific conditions of the site. Extra protection such as riprap should be provided as needed to control potential erosion.

Mowing with the proper equipment to handle rights-of-way vegetation has proven to be one of the most economical methods for vegetation control. By restricting tall-growing plants, better line-of-sight is provided as a safety factor. Urban areas and approaches to towns and cities are mowed with greater frequency for aesthetic reasons, but this adds to the cost. To be successful, mowing operations should be planned and monitored for the specific vegetation to be controlled.[12]

Controlled burning, with or without mowing, is another management tool that is efficient and economical. It must be properly managed for the safety of both operators and the public. Care must be exercised to prevent fires from expanding onto adjacent property, to prevent the smoke from becoming a hazard to transport vehicles, and to protect desirable vegetation. A permit is often required so that fire-fighting equipment is available.

Nonselective Herbicide Control

Nonselective herbicides are chemicals which kill all vegetation and leave the ground sterile and barren for a period of time. This is often desirable in certain situations, such as around the poles of power lines, ballast of railroad beds, and the borders of paved highways. The major problem with such applications is the potential for erosion. To prevent erosion, the treatment area may be capped with asphalt or the chemical may be mixed with asphalt and applied as a treatment.

Table 1
SOME COMMON USES OF SELECTED HERBICIDES[14]

Herbicide	Action[a]	Uses	Special values and problems
Ammonium sulfamate (ammate)	HGR	Stem-foliar application; treating stumps	Especially effective in control poison ivy, poison oak, and poison sumac, safe for use around water and crops
Trichloro-benzoic acid (TBA)	HGR	Often used to increase spectrum of kill as additive	Spray drift may injure adjacent crops
Bromacil (Hyvar®)	HGR	Basal treatment in a water solution	Leaves drop quickly after treatment, so brownout problems are eliminated; may persist in soil
Dicamba (Banvel®)	HGR	Often used in combination with other herbicides; controls sprouting of stumps and stubble	Effective on hard-to-kill species, including ash, gum, maple, pine, sourwood, mesquite, witch-hazel, creosote bush, sagebrush, and yaupon; selective on grass
Karbutilate®	HGR	Basal treatment in a water solution; controls sprouting of stumps and stubble	Persists in soil
Picloram (Tordon®)	HGR	Stem-foliar application; basal application; treating stumps; used mainly in combination with other herbicides; effects are usually additive	Effective against many woody plants; delrayed killing action; persists in soil; various mixtures
2,4-D	HGR	Stem-foliar application	Kills both woody and herbaceous vegetation; selective on grass
2,4,5-T	HGR	Stem-foliar application	More effective woody plant killer than 2,4-D; selective on grass
2,4-DP	HGR	Stem-foliar application; controls root-suckering plant species	
Silvex® (2,4,5-TP)	HGR	Stem-foliar application	Controls some species, including maple and oak, that are resistant to 2,4-D and 2,4,5-T; persists in soil
MCPA (2-methyl-4-chlorophen-oxyacetic acid)	HGR	Stem-foliar application	Safe around croplands; when added to other herbicides, gives wider spectrum control

[a] HGR, hormone growth regulator.

Selective Herbicide Control

Selective herbicides (see Table 1) are used to control specific undesirable vegetation. The removal of such vegetation improves the right-of-way environment and will have a beneficial effect for the well-being of the transport facility, its uses, and adjacent property owners. To be successful, the weedy vegetation must be correctly identified, the right chemical applied at the right time, and care used to avoid undesirable effects to adjacent property owners. Wherever possible, spot treatments are preferred to minimize the hazard to desirable plants.[12]

Integrated Control

Vegetation in rights-of-way areas can be controlled by selective removal of plants, either by mechanical means or by chemical herbicides. This requires intensive management to maintain and/or restore the quality of the existing cover. Plants found to be objectionable under normal circumstances, including tall growing trees, poisonous plants, and those species not wanted, may be removed by local application of herbicide or by hand removal. Mechanical cutting is then applied as a regular program at proper intervals. In general, there is a steady decline of undesirable species, and existing low-growing species will persist.[13]

The process of attaining a full canopy under this program is relatively slow. The precise estimate of time required to achieve this state cannot be easily defined, but with good starting density of native plants, a good balance can usually be obtained in 5 to 10 years. This form of integrated management has been found to be less expensive than traditional methods, and it fosters a diversified and esthetically pleasing natural product that will supply important cover and food niches for wildlife of the area.[13]

MANAGEMENT OF RIGHTS-OF-WAY FOR FISH AND WILDLIFE

A variety of techniques exist for managing rights-of-way for fish and wildlife populations. The right-of-way is unique in that it provides cover with an ''edge effect'' that is especially conducive to wildlife development and persistence. This permits each species to find its best habitat in the variety of vegetation that exists at the edge of a tall tree canopy.[14]

Management techniques to completely eliminate vegetation can be very harmful to a wildlife population, therefore, selective removal should be practiced. Brush piling can be used to provide additional cover in this operation. It also provides a variety of cover as the grasses and understory begin regrowth. Selective herbicide application is the most often used technique on rights-of-way where wildlife protection is mandatory. If practiced with full knowledge of the canopy intended, it can be used to facilitate production of good natural habitats for wildlife. While there are few rights-of-way that involve fish habitat, the same general principles apply to this vegetation management. Moreover, there is no single management technique that can be totally used, therefore, combinations of methods should be used to provide the most satisfactory result.

SUMMARY AND CONCLUSIONS

Programs of rights-of-way vegetation control are an essential part of good utilization and management. They must be carefully planned and executed to obtain the desired effects. This includes required control features of the right-of-way, esthetic setting of the operation, fish and wildlife cover, general conditions of safety, etc. The actual program must also be cost effective. Nevertheless, it is possible to have proper vegetation control without undue cost or inconvenience.

REFERENCES

1. **Sylvester, E. P.**, Control of roadside weeds and brush, *Weeds*, 1, 17, 1951.
2. **Zukel, J. W. and Eddy, C. O.**, Present use of herbicides in highway area, *Weeds*, 6, 61, 1958.
3. **Parker, W. B.**, Weed and brush control on railroads, *Proc. South. Weed Conf.*, 18, 346, 1965.
4. **Yazell, D. H.**, Problems encountered on railroad rights-of-way, *Proc. North Cent. Weed Control Conf.*, 22, 40, 1965.
5. **Lowry, J. M.**, Programmed weed control on the cotton belt, *Proc. South. Weed Conf.*, 18, 359, 1965.
6. **Francisco, D. C. and Aldred, J. R.**, The control of undesirable vegetation on TVA's steam and hydro plant areas, *Proc. South. Weed Conf.*, 16, 299, 1963.
7. **Bell, T. O.**, Pipeline public relations and their effect on chemical applications to rights-of-way, *Proc. South. Weed Conf.*, 18, 378, 1965.
8. **Hubbard, C. C.**, Humble's experience with soil sterilants in southeast Texas, *Proc. South. Weed Conf.*, 18, 430, 1965.
9. *Farm Chemicals, Pesticide Dictionary*, Meister, Willoughby, Ohio, 1975.
10. **Bovey, R. W.**, Response of Selected Woody Plants in the United States of Herbicides, Agriculture Handbook No. 493, U.S. Government Printing Office, Washington, D.C., 1975.
11. **Spurrier, E. C.**, Glycosate — a new broad spectrum herbicide, *PANS, (Pest Artic. News Summ.)*, 19, 607, 1973.
12. **Niering, W. A. and Goodwin, R. H.**, Creation of relatively stable shrublands with herbicides, arresting "succession" on rights-of-way and pastureland, *Ecology*, 55(4), 784, 1974.
13. **Rusch, A. J., Thompson, D. R., and Kabat, C.**, Management of Roadside Vegetation Cover by Selective Control of Undesirable Vegetation, Tech. Bull. 117, Department of Natural Resources, Madison, Wis., 1980.
14. Fish and Wildlife Service, Management of Transmission Line Rights-of-way for Fish and Wildlife, FWS/OBS-79/22, U.S. Department of the Interior, Washington, D.C., 1979.

Chapter 13

FACTORS AFFECTING COMMERCIAL APPLICATION OF HERBICIDES

INTRODUCTION

The maintenance of rights-of-way of large utility transmission systems, some of which are in excess of 20,000 km in length, presents a formidable problem. One of the most economical approaches for herbicide application is the use of rotary-wing aircraft or helicopters. Helicopters have an advantage over fixed-wing aircraft in that they are more maneuverable.

Apart from government regulation and public acceptance, the major restriction on aircraft spraying is suitable weather. Most operators with liquid spray units do not fly if the wind velocity is over 7 km/ha. Some operators feel that helicopters are more satisfactory than fixed-wing aircraft because a lower flying speed is possible and greater maneuverability of the helicopter contributes to more accurate placement of the spray pattern.

FOLIAR SPRAY OPERATIONS

Grundy and Bennett[1] have pointed out that 70 to 90% topkill is obtained on poplar, birch, willow, and tag alder with applications of 4 kg of acid equivalent per 30 ℓ of a formulation of 2,4-dichlorophenoxy (2,4-D) and 2,4,5-trichlorophenoxy (2,4,5-T) in oil. The conifers, balsam, fir and spruce, are resistant to 2,4-D and 2,4,5-T sprays. Their work with trichloroacetic acid (TCA) was satisfactory; rates of about 25 kg of acid equivalent per hectare gave good control over parts of the right of way. They used two 11-ft booms for application on 100-ft rights-of-way. Control was confined in most instances to a strip on each side of the power line directly under the path of the helicopter.

PELLET APPLICATIONS

Pellets are an approach to the drift problem in aerial spraying.[2-5] Some of the advantages of using pellets are:

1. Material can be applied in winds up to 12 mi/hr.
2. Brownout is eliminated
3. The material is nonvolatile.
4. Helicopters can be used for pellet application before the following spray season.

The disadvantages are

1. Cost
2. Some tendency to kill brush off of the rights-of-way because of the extension of the lateral root systems
3. Adequate rainfall needed to move the material into the ground

According to some tests, the materials (Dybar® and Tordon® 10K) are highly satisfactory for brush control at various rates.[2-5]

MIXED APPLICATIONS

A problem occurred on the Tennessee Valley Authority (TVA) study[3] in that many tracts of land or farms were cleared several months before line construction was completed. The interval resulted in a medium to dense stand of brush ranging from 2 to 5 m in height on the right-of-way at the time of the line was energized. The stem count in the area ranged from about 2000 to 5000/ha and consisted of ash, cedars, dogwood hickory, maples, oaks, persimmons, pine, poplar, sassafras, sourwood, and sumac. The application was made by knapsack sprayers because of the rough mountainous terrain. The stump-treated area did not require treatment until after three full growing seasons, and then it was foliage treated with ammate near crops and with esters in other areas.

RETREATMENT

After foliar applications, it is usually necessary within a few years to use a different approach to brush control to eliminate the more resistant species (for example, ash, maple, and elm). Usually, this is done by basal or dormant treatment. The proper spraying method and herbicide section through rights-of-way surveys of the species present, their height, and stem count or density is necessary. It has been calculated that the stem count for basal or dormant application, to be economically feasible, should average 1000/ha or less. One formulation for typical dormant spraying would consist of 12:1, of 2,4,5-T esters (4 lb acid per gallon and 97 lb of diesel oil). Application is made to the brush within a foot of the ground line, thus permitting the mixture to run down and thoroughly wet the root collar.

In many areas of this country and Canada where utility rights-of-way pass through pine forests, conifers become a serious maintenance problem. Herbicides 2,4-D and 2,4,5-T are effective in controlling deciduous woody growth, but have little effect on the conifers; in fact, 2,4,5-T is the herbicide of choice for pine release. The conifers grow much more rapidly than normal when the competition is reduced for light, moisture, and nutrients, and they become a dominant vegetation in rights of way. Oil-borne sprays containing 2,4-D and 2,4,5-T have been used for conifer control in the northeastern states. The extensive use of oil for spraying under Canadian conditions is limited mainly by the cost and difficulty of transporting the required oil volume into remote areas. Effective control of conifer, balsam, fir, and black spruce can be obtained with water-borne sprays of monuron, 2,3,6-trichlorobenzoic acid (2,3,6-TBA), Silvex®, TCA, and dalapon applied during the growing season.

PROGRAM PLAN

Telephone lines are special types of rights-of-way as they are usually associated with roadways. The maintenance of the line in terms of vegetation control involves a wide degree of planning. Coffey[6] indicates that the regional cutting intervals are 5 to 10 years in the Rocky Mountain states, 3 to 5 years in the southwest, and 1, 2, or 3 years in most of the other regions of the U.S. Coffey has outlined the planning and programming for conversion from brush cutting of rights of way to chemical control. He indicates that good management requires the developing of a long-range plan for chemical control and believes that the major steps in planning are

1. Brush inventory
2. Job specifications
3. Material supervision
4. Field inspection

PERSISTENCE OF HERBICIDE RESIDUES

In support of programs testing aerial dissemination systems, a 1-mi^2 test grid on Eglin Air Force Base Reservation, Fla. received massive quantities of herbicides.[26] The purpose of these test programs was to evaluate the capabilities of the equipment systems, not the biological effectiveness of the various herbicides. Hence, it was only after repetitive applications that test personnel began to express concern over the potential ecological and environmental hazards that might be associated with continuance of the Test Program. This concern led to the establishment of a research program to measure the ecological effects produced by the various herbicides on the plant and animal communities of the Test Area. This report documents 6 years of research[27,28] on the Test Area and the immediately adjacent streams and forested areas.

From the high rates of herbicides that were applied during the years of testing spray equipment, it was obvious that the Test Area offered a unique opportunity to study herbicide persistence and soil leaching. Yet the problem of how best to assess the level of herbicide residue was a difficult one. The herbicides could be chemically present, but because of soil binding might not be biologically active. Thus, both bioassay techniques and analytical analyses were employed. The first major bioassay experiment was conducted in April 1970. By considering the flightpaths, the water sources, and the terracing effects, it is possible to divide the 1-mi^2 test grid into 16 vegetation areas. These areas formed the basis for the random selection of 48 3-ft soil cores. Soybean bioassays indicated that 27 of the 48 cores were significantly different from control cores (95% probability level). The results indicated that soil leaching or penetration was much more prevalent along the dissemination flight paths than in other areas of the test grid. Efforts to quantitate (chemically) the bioassay were confined to only the top 6-in. increment because of within-core variations. By considering that all phytotoxic effects were from herbicide orange (2,4-D and 2,4,5-T), the average value for the top 6 in. of soil core for the eight cores showing greatest herbicide concentration was 2.82 ppm herbicide. Chemical analyses of soil cores collected from the eight sites showing greatest phytotoxic concentrations were performed in December 1970. Results indicated that the maximum concentration of either 2,4-D or 2,4,5-T was 8.7 ppb.

EFFECTS ON VERTEBRATE ANIMALS

A total of 73 species of vertebrate animals (mammals, birds, reptiles, and amphibians) were observed on Test Area C-52A and in the surrounding area.[27] Of these 73 species, 22 species were observed only off the grid, 11 species were observed only on the grid, and 40 species were observed to be common to both areas. During the early studies, no attempts were made to quantitate animal populations in the areas surrounding the grid; however, in 1970, preliminary population studies by trap-retrap methods were performed on the beach mouse (*Peromyscus polionotus*) population for a 60-day period to confirm the hypothesis that it was the most prevalent species on the grid. The hypothesis was supported by the capture of 36 beach mice from widely distributed areas on the grid, except in areas with less than 5% vegetation. Eight pairs of eastern harvest mice were taken to the laboratory and allowed to breed. Six of the eight pairs had litters totalling 24 mice. These progeny were free from any gross external birth defects. During February to May 1971, population densities of the beach mouse were studied at eight different locations on the grid along with two different areas off the grid which served as controls. Populations were estimated on the basis of trap-retrap data. There was no difference in mouse population densities in herbicide-treated and control areas affording comparable habitats. All indications were that any population differences in other animal species between the test area and the surrounding area were due to differences caused by the elimination of certain plants and, therefore, certain ecological niches, rather than being due to any direct detrimental effects of the herbicides on the animal population present.

Nine mice were captured and taken to the laboratory for postmortem pathological examination. There were no instances of cleft palate or other deformities. Histologically, liver, kidney, and gonadal tissues from these animals appeared normal. Several other species of animals were caught, both on and off the test grid. These included beach mice (*P. polionotus*), cotton mice (*P. gossypinus*), eastern harvest mice (*Reithrodontomys humulis*), hispid cotton rats (*Signodon hispidus*), six-lined race-runners (*Cnemidophorus sexlineatus*), a toad (*Bufo americanus*), and a cottonmouth water moccasin (*Ancistrodon piscivorus*). A total of 89 animals were submitted to The Armed Forces Institute of Pathology, Washington, D.C. for complete pathological examination, including gross and microscopic studies. Liver and fat tissue from 70 rodents were forwarded to the Interpretive Analytical Services, Dow Chemical, Midland, Mich. for tetrachlorodibenzo-*p*-dioxin (TCDD) analyses. The sex distribution of the trapped animals was relatively equal. The ages of the animals varied, but adults predominated in the sample. No gross or histological developmental defects were seen in any of the animals. Several of the rats and mice from both groups were pregnant at the time of autopsy. The stage of gestation varied considerably from early pregnancy to near term. The embryos and fetuses were examined grossly and microscopically, but no developmental defects or other lesions were observed. Gross necropy lesions were relatively infrequent and consisted primarily of lung congestion to those animals that had died from heat exhaustion prior to being brought to the laboratory. The organ weighs did not vary significantly between the test and control animals when an animal with lungs and kidneys showing inflammatory pathological lesions was removed from the sample. Histologically, the tissues of 13 of the 26 control animals and 40 to 63 animals from the test grid were considered normal. Microscopic lesions were noted in some animals from both groups. For the most part, these were minor changes of a type one expects to find in any animal population.

Beach mice were not found on the more barren sections of the grid (0 to 5% vegetative cover). There were, however, some areas of the grid which had population densities exceeding those of the species preferred habitat as reported in the literature. In an attempt to correlate distribution of the beach mouse with with vegetative cover (i.e., habitat preferance), a trapping-retrapping program of 8 days duration was conducted. The majority of animals[63] were found in areas with 5 to 60% vegetative cover; within this range, the greatest number of animals trapped[28] was from an area with 40 to 60% cover. A similar habitat preference has been observed along the beaches of the Gulf Coast.

Trapping data was collected during the 1971 to 1973 period to determine whether an increase in the population of beach mice had occurred. The statistical evidence derived from that study showed that the population increased slightly. The population of beach mice was also higher in 1973 than in 1971 in the area of the test grid. The apparent increase in beach mouse population on the grid in 1973 over 1971 was probably due to the natural recovery phenomenon of a previously disturbed area (i.e., ecological succession). Some areas of the test grid have currently exceeded that preferred percentage of vegetative coverage of the beach mouse habitat, and other areas were either ideal or fast developing into an ideal habitat.

A 1973 sweep net survey of the arthropods of the Test Area resulted in the collection of over 1700 specimens belonging to 66 insect families and arachnid orders. These totals represented only one of five paired sweeps taken over a 1-mi section of the test grid. A similar study performed in 1971 produces 1803 specimens and 74 families from five paired sweeps of the same area using the same basic sampling techniques. A much greater number of small to minute insects were taken in the 1973 survey. Vegetative coverage of the test area had increased since 1971. The two studies showed similarities in pattern of distribution of arthropods in relation to the vegetation, number of arthropod species, and arthropod diversity. Generally, the 1973 study showed a reduction of the extremes found in the above parameters in the 1971 study.

There are two classes of aquatic areas associated with the Test Area: ponds actually on the square mile area and streams which drain the area. Most of the ponds are primarily of the "wet weather" type, drying up once in the last 5 years, although one of the ponds is spring fed. Three major streams and two minor streams drain the test area. The combined annual flow of the five streams exceeds 24 billion gal of water. Seventeen different species of fishes have been collected from the major streams, while three species have been collected from the spring-fed pond on the grid. Statistical comparisons of 1969 and 1973 data of fish populations in the three major streams confirm a chronologically higher diversity in fish populations. However, the two control streams confirm a similar trend in diversity.

TCDD analysis of biological organisms from streams draining the Test Area or in the ponds on the Test Area were free from contamination at a detection limit or less than 10 parts per trillion. In analyses performed 3 years after the last application of 2,4-D and 2,4,5-T herbicide, the test grid exhibited population levels of soil microorganisms identical to that in adjacent control areas of similar soil and vegetative characteristics not exposed to herbicides. There were increases in actinomycete and bacterial populations in some test site areas over levels recorded in 1970. This was possibly due to a general increase in vegetative cover for those sampling sites and for the entire test grid. No significant permanent effects could be attributed to exposure to herbicides.

2,4,5-T AND TCDD

Concern over the level of contamination 2,4,5-T herbicide by the teratogen 2,3,7,8-TCDD may result in the use and/or disposal of these herbicides has been discussed. The premise is made that high concentrations of phenoxy herbicides and TCDD will degrade to innocuous products by combined action of soil microorganisms and soil chemical hydrolysis.[10-17, 20-24]

In August 1972, a site for the soil incorporation of phenoxy herbicides was selected on the Air Force Logistics Test Range Complex at Hill Air Force Base, Utah.[27] The potential site was characterized as being relatively flat, with a uniform surface without crops of rocks or other distorting formations. Soil samples were taken 1 day following application and at specified days after application. In all cases, soil cores were taken in the center of the trench, and the herbicide analysis was made by the standard method of gas chromatographic analysis.

HERBICIDE DEGRADATION

The results of the analysis of soil samples taken from the test plots are given in Tables 1 to 3. Table 1 illustrates the loss of total active herbicide from the upper 12 in. of soil. The percent loss over a 7-month sampling period was 78, 75, and 61% for 1000, 2000, and 4000 lb active ingredient (ai) per acre, respectively. If strict half-life exponential decay is assumed, the half-life ranges from 146 to 155 days.

Tables 2 and 3 illustrate the individual losses of 2,4-D and 2,4,5-T contained in the original formulation. Since the TCDD concentration of the formulation was known, and its determination in the soil core was performed by the same laboratory, an estimation can be made of the degradation of TCDD by a comparison to the expected values of herbicide at the time of sampling. The percent loss of TCDD over a period

Table 1
SOIL CONCENTRATION, PARTS PER MILLION, OF 2.4-D AND 2,4,5-T HERBICIDE AT SELECTED SAMPLING PERIODS, DAYS, FOLLOWING SOIL INCORPORATION

Application rate of formulation (lb ai/acre)[a]	Sampling time after incorporation (days)[b]				
	0[c]	110	220	282	440
1,000	10,000	5,580	1,876	—	1,216
2,000	20,000	11,877	—	4,670	2,944
4,000	40,000	17,729	—	8,489	6,944

[a] Pounds active ingredient per acre.
[b] Data represent an average of two replications with the duplicate samples of each replication: the total value for depths 0 to 6 and 6 to 12 in. of soil increment.
[c] Theoretical concentration at time of application based on a 2-in. spray swath at a depth of 4 to 6 in. within the soil profile.

Table 2
SOIL CONCENTRATION, PARTS PER MILLION, OF 2,4-D HERBICIDE AT SELECTED SAMPLING PERIODS, DAYS, FOLLOWING SOIL INCORPORATION

Approximate rate of 2,4-D[a] (lb ai/acre)	Sampling time after incorporation (days)[b]				
	0[c]	110	220	282	440
500	5,000	3,280	976	—	616
1,000	10,000	7,261	—	2,370	1,844
2,000	20,000	10,545	4,829	—	4,112

[a] Rate of herbicide per acre was based on the original specification of the formulation (i.e., and 50:50 *n*-butyl formulation containing 8.63 lb active ingredient per gallon).
[b] Data represent an average of two replications with two duplicate samples of each replication: the total for depths 0 to 6 and 6 to 12 in. of soil increment.
[c] Theoretical concentration at time of application based on 2-in. spray swath at a depth of 4 to 6 in. within the soil profile.

of 265 days was 87.7%. Therefore, the estimate half-life of TCDD is 88 days, under the conditions described.

ENVIRONMENTAL ASSESSMENT OF TCDD

Detection of the highly potent impurity TCDD in the herbicide 2,4,5-T necessitated an environmental assessment of the impact of this contaminate. Information was needed on movement, persistence, and plant uptake to determine whether low concentrations reaching plants, soils, and water posed any threat to man and his environment. Synthesis of uniformly labeled[14] C-TCDD greatly facilitated detection of TCDD in soil and plant experiments. For unlabeled experiments, it seemed prudent to use only small

Table 3
SOIL CONCENTRATION, PARTS PER MILLION, OF 2,4,5-T HERBICIDE AT SELECTED SAMPLING PERIODS, DAYS, FOLLOWING SOIL INCORPORATION

Approximate rate of 2,4,5-T[a] (lb ai/acre)	Sampling time, after incorporation (days)[b]				
	0[c]	110	220	282	440
500	5,000	2,300	900		604
1,000	10,000	4,616		2,300	1,100
2,000	20,000	7,184		3,734	2,832

[a] Rate of herbicide per acre was based on the original specification of the formulation (i.e., a 50:50 *n*-butyl formulation containing 8.63 lb active ingredient per gallon).
[b] Data represent an average of two replications with two duplicate samples of each replication: the total for depths 0 to 6 and 6 to 12 in. of soil increment.
[c] Theoretical concentration at time of application based on 2-in. spray swath at a depth of 4 to 6 in. within the soil profile.

quantities of diluted solutions in situations where decontamination was feasible and to rely on the sensitivity afforded by electron capture chromatography.[8-10,12,21,23,25,26] In addition to the movement, persistence, and plant uptake studies, it was essential to determine whether chlorinated phenols could be condensed to their corresponding chlorodioxins by photochemical or biochemical processes under natural conditions.

The mobilities of two chlorinated dioxins, 2,7-dichlorodibenzo-*p*-dioxin (DCDD) and TCDD, and 2,4,5-T were evaluated by the soil TLC technique. Radioisotopically labeled compounds are spotted at the origin, and the plate is developed with water, dried, and covered with "no screen" X-ray film to locate the compound. Five soils varying widely in texture and other properties were tested: Norfolk and Lakeland sandy loams, Hagerstown silty clay loam, Barnes clay loam, and Celeryville muck.

Both dioxins were found to be immobile in all soils, i.e., they would not be leached into the soil by rainfall or irrigation. The relative immobility of the chlorodioxins is not unexpected based on the very low solubility of these compounds in water (0.6 μg/ℓ). In contrast, the herbicide 2,4,5-T is relatively mobile in sandy soils, but movement decreases as soil organic matter increases.

Soil persistence experiments were undertaken

1. To learn the effect of soil type and concentration on the rate of TCDD degradation
2. To isolate and characterize degradation products from DCDD and for TCDD
3. To determine whether chlorodioxins could be formed from condensation of chlorophenols in the soil environment

No evidence of DCDD or TCDD could be found in soils receiving 10, 100, and 1000 ppm of the 2,4-dichlorophenol (DCP) or the 2,4,5-trichlorophenol (2,4,5-TCP) as determined by electron capture gas chromatography.

To determine the persistence of chlorodioxins, concentration rates of 1, 10, and 100 ppm of unlabeled TCDD were established in soils and then assayed periodically for dioxin residues.

Measurements on dioxin residues after 20, 40, 80, 160, and 350 days indicated a relatively slow degradation process in both soils. At 350 days, 56% of the initially applied TCDD was recovered from the sandy soil, and 63% was recovered from the silty clay loam for all concentrations. A second experiment was conducted using ^{14}C-TCDD at 1.78, 3.56, and 17.8 ppm in the same two soils. After 350 days, the soils were combusted, and $^{14}CO_2$ trapped and determined by liquid scintillation. The amount recovered from Lakeland was 67% at 1.78 ppm, 70% at 3.56, and 73% at 17.8 ppm. For Hagerstown the recoveries were 52, 52, and 89% over the same concentrations.

Plant uptake is one of several routes by which an organic contaminant can enter man's food chain.[11,14-19,24] The amount of uptake depends on plant species, concentration, depth of placement, soil type, temperature, moisture, and a host of other parameters. Translocation of the absorbed material into various plant parts will determine the degree of man's exposure, i.e., whether the material moves to an edible portion of the plant. Past experience with nonpolar chlorinated pesticides suggested optimal uptake conditions are achieved when the chemical is placed in a soil with low adsorptive capacity (e.g., a sand), evenly distributed throughout the soil profile and with oil producing plants. Plant experiments were conducted with one set of parameters that would be optimal for uptake and translocation. The uptake of two dioxins and one phenol 2,4-DCP from one soil was measured in soybean and oats. The rates of application were DCP = 0.07 ppm, DCDD = 0.10 ppm, and TCDD = 0.06 ppm. Oats and soybeans (at all growth stages) accumulated very small quantities of chlorodioxins or the chlorophenol. A maximum of 0.21% of the phenol and 0.15% of the dioxin present in soils was translocated to the aerial portion of oats and soybeans. No detectable amounts of ^{14}C were found in the grain or soybeans harvested at maturity. The amount of TCDD applied to these soils would be many thousand times greater than that which would occur in soils from herbicide applications containing a few parts per million of TCDD as an impurity.

To deal with the question of translocation of foliar-applied TCDD, labeled dioxins were applied to the center leaflet of the first trifoliate leaf of 3-week-old soybean plants. All compounds were applied in an aqueous surfactant solution (Tween 80®) to enhance leaf adsorption and to keep the water-insoluble dioxins in solution. Plants were harvested 2, 7, 14, and 21 days after treatment, dissected into treated and untreated parts, and analyzed separately. Neither dioxin nor chlorophenol was translocated from the treated leaf. A rapid loss of the dichlorodioxin and dichlorophenol occurred from the leaf surface. This loss may have been due to volatilization. Very little TCDD was lost from soybean leaves, while a gradual loss (38% in 21 days) did occur from oat leaves.

The following facts have emerged from studies with 2,7-DCDD and 2,3,7,8-TCDD:

1. They are not biosynthesized by condensation of chlorophenols in soils.
2. They are not photoproducts of 2,4-DCP.
3. They do not leach into the soil profile and consequently pose no threat to groundwater.
4. They are not taken up by plants from minute residues likely to occur in soils.
5. Photodecomposition is insignificant on dry soil surfaces but is probably important in water.
6. Dichlorodibenzo-*p*-dioxin is lost by volatilization, but TCDD is probably involatile.
7. They are not translocated within the plant from foliar application.

SUMMARY AND CONCLUSIONS

The results of these and many other studies indicate commercial application of herbicides (2,4-D and 2,4,5-T in particular) is effective, and they do not accumulate in any of the components of the biosphere, nor do they accumulate in the tissues of animal products used for human consumption. There is no indication that TCDD accumulates in air, water, or plants, even though it may remain active for some time in the soil after extremely heavy rates of application.

REFERENCES

1. **Grundy, W. M. and J. M. Bennett,** Aerial application of herbicides for right-of-way brush control, *Down Earth,* 16 (2), 6, 1960.
2. **Mann, R. A.,** Stump treatment after initial clearing on right-of-way of newly constructed lines, *Proc. South. Weed Conf.,* 16, 318, 1963.
3. **Mann, R. A.,** Pellet application by helicopter on TVA rights-of-way, *Proc. South. Weed Conf.,* 19, 307, 1966.
4. **Mann, R. A. and J. R. Aldred,** Brush control of utility rights-of-way by dormant application, *Proc. South. Weed Conf.,* 17, 295, 1964.
5. **Nation, H. A.,** Woody plant control on utility rights-of-way with "Tordon" herbicide pellets, *Proc. South. Weed Conf.,* 18, 387, 1965.
6. **Coffey, W. J.,** Planning and programming and brush control, *Down Earth,* 11 (1), 10, 1955.
7. **Courtney, K. D. and Moore, J. A.,** *Teratology* studies with 2,4,5-trichlorophenoxyacetic acid and 2,3,7,8-tetrachlorodibenzo-*p*-dioxin, *Toxicol. Appl. Pharmacol.,* 20, 396, 1971.
8. **Crosby, D. G. and Wong, A. S.,** Environmental degradation of 2,3,7,8-dibenzo-*p*-dioxin (TCDD), *Science,* 195, 1337, 1977.
9. **Emerson, J. L., Thompson, D. J., Strebing, R. J., Gerbig, C. G., and Robinson, V. B.,** Teratogenic study of 2,4,5-trichlorophenoxyacetic acid in the rat and rabbit, *Food Cosmet. Toxicol.,* 9, 395, 1971.
10. **Gehring, P. J., Kramer, C. G., Schwetz, B. A., Rose, J. Q., and Rowe, J. K.,** The fate of 2,4,5-trichlorophenoxyacetic acid (2,4,5-T) following oral administration, *Toxicol. Appl. Pharmacol.,* 26, 352, 1973.
11. **House, W. B., Goodson, L. H., Gadberry, H. M., and Dockter, K. W.,** Assessment of Ecological Effects of Extensive or Repeated use of Herbicides, Final Report, AD824, Midwest Research Institute, Kansas City, Mo., 1967, 314.
12. **Kearney, P. C., Woolson, E. A., and Ellington, C. P., Jr.,** Persistence and metabolism of chlorodioxins in soils, *Environ. Sci. Technol.,* 6, 1017, 1972.
13. **Khanna, S. and Fang, S. C.,** Metabolism of C^{14}-labeled 2,4-dichlorophenoxyacetic acid in rats, *J. Agric. Food Chem.,* 14, 500, 1966.
14. **Kirch, J. H.,** Special herbicide combination for rights-of-way brush control, in *Symp. Proc. Herbicides and Vegetation Management,* Oregon State University, Corvallis, 1967.
15. **Klingman, D. L. and Shaw, W. C.,** Using Phenoxy Herbicides Effectively, U.S. Department of Agriculture, Farmer's Bull. No. 2183, U.S. Government Printing Office, Washington, D.C., 1967.
16. **Lutz, J. F., Byers, G. E., and Sheets, T. J.,** The persistence and movement of picloram and 2,4,5-T in soils, *J. Environ. Qual.,* 2, 485, 1973.
17. **Mitchell, J. W., Hodgson, R.E. and Gaetjena, C. F.,** Tolerance of farm animals to feed containing 2,4-D, *J. Anim. Sci.,* 5, 226, 1946.
18. **Palm, C. E., Ed.,** *Principles of Plant and Animal Pest Control,* Vol. 2, Publ. 1597, National Academy of Sciences, Washington, D. C., 1968.
19. **Pimentel, D.,** Ecological Effects of Pesticides on Non-Target Species, U.S. Government Printing Office, Washington, D.C., 1971.
20. **Schultz, D. P.,** Dynamics of a salt of (2,4-dichlorophenoxy) acetic acid in fish, water and hydrosol, *J. Agric. Food Chem.,* 21, 186, 1973.
21. **Schwetz, B. A.,** Toxicology of chlorinated dibenzo-*p*-dioxins, *Environ. Health Perspect.,* 5, 87, 1973.
22. **Smith, A. E.,** The hydrolysis of 2,4-dichlorophenoxyacetate esters to 2,4-dichlorophenoxyacetid acid in Saskatchewan soils, *Weed Res.,* 12, 364, 1972.

23. **Sterling, T. D.**, Difficulty of evaluating the toxicity and teratogenicity of 2,4,5-T from existing animal experiments, *Science,* 174, 1358, 1971.
24. U.S. Department of Agriculture, Report and Tables on Domestic Use of a 2,4,5-T, 1964 and 1966, Production Resources Branch, Farm Production Economics Division, U.S. Department of Agriculture, Washington, D.C., 1970.
25. **Woolson, E. A. Thomas, R. F., and Ensov, P. D. J.**, Survey of polychlorodibenzo-*p*-dioxin content in selected pesticides, *J. Agric. Food Chem.,* 20, 351, 1972.
26. **Young, A. L., Ed.**, Ecological Studies on a Herbicide-Equipment Test Area Eglin AFB Reservation, Fla., AFATL-TR-74-12, Eglin Air Force Base Reservation, Fla., 1974.
27. **Young, A. L., Arnold, E. L., and Wachinski, A. M.**, The ecological consequences of massive quantities of 2,4-D and 2,4,5-T herbicides — summary of a five year field study, paper presented to the Weed Science Society of America, Las Vegas, Nev., 1974.

Chapter 14

TOXICOLOGICAL SIGNIFICANCE OF RIGHTS-OF-WAY CONTROL

INTRODUCTION

Although herbicides in general are toxic to plants and nontoxic to humans, some are toxic to humans and must be handled with care. Like all pesticides, they may enter the body in three ways, i.e., the mouth, the skin, and the nose. Each pesticide chemical therefore has three different toxicities, depending on the route of entry.

There have been many scientific studies to determine the behavior of herbicides in plant and animal systems under varied environmental conditions. It is important to note that the esters of herbicides can hydrolyze. Thus, the behavior of the pure acids and their salts are also pertinent and will be discussed in the following paragraphs along with characteristics of ester forms.

The differences in toxic effects produced by the various salts, amines, and esters can often be explained on a pharmacokinetic basis in which the concentrations at the receptor sites in the organism depends on the absorption and distribution rates in relation to the rates of metabolism and excretion. The rate of absorption into plants or animals will be dependent upon various interrelated factors, such as route of entry and rate of membrane transport. Specific membrane transport rate will depend upon the characteristics of the membrane in relation to the size, shape, polarity, and lipid solubility of the particular herbicide molecule.

Oral Toxicity

Herbicides may enter the body through the mouth. They may be taken in by mistakes while eating or smoking on the job or when improperly stored in food containers. Sometimes herbicides are swallowed on purpose as in suicide. The most common route of accidental absorption in terrestrial animals is via ingestion. This is especially true in herbivores. However, absorption of toxic doses via inhalation and cutaneous routes is possible, if uncommon. The literature indicates that gastric absorption of 2,4-dichlorophenoxy (2,4-D) and 2,4,5-trichlorophenoxy (2,4,5-T) and their amines and alkali salts occur readily as would be predicted from classical Henderson-Hasselbalch relationships. However, the (GI) absorption of 2,4-D in the form of an ester may be incomplete. Erne[3] administered 2,4,-D ester orally and found no detectable esters in the plasma. However, detection of low levels of 2,4,-D in the plasma indicated that some hydrolysis of the ester had occurred. In studies with rats, calves, chickens, and pigs,[4] he found that the highest tissue levels of 2,4-D and 2,4,5,-T were in liver, kidney, lung, and spleen, levels sometimes exceeding the plasma level. In blood cells, 10 to 20% of the plasma level was found. Penetration of 2,4-D into adipose tissue and into the central nervous system (CNS) was restricted, whereas a ready placental transfer was demonstrated in swine. The distribution pattern did not show any significant species or sex differences.

Dermal Toxicity

Pesticides can also be taken in through the skin. Wearing clothing that is wet with pesticides or allowing pesticides to fall directly on the skin while mixing or spraying can be dangerous. Dry material dusts, wettable powders, or granules can also be absorbed. Some pesticides do not pass through the skin very quickly. Others, however, can be as dangerous when entering through the skin as when taken by mouth.

Pesticides pass through the skin on some areas of the body more quickly than on other areas. You must be especially careful with your eyes. The back of the hands and the wrists absorb more than the palms do. The armpits, back of the neck, groin, and feet take in pesticides easily too. Cuts or scrapes allow pesticides to enter even more easily than usual.

Inhalation Toxicity

You can breathe in pesticides too. Pesticides in the form of dusts, spray mist, or fumes can be sucked into your lungs as you inhale. This route of entry is important during mixing of wettable powders, dusts, or granules. You can also be poisoned when fumigating or spraying without the proper respirator in a closed area, such as greenhouses, apartments, or grain bins. Even inhalation of dilute pesticides can result in poisoning.

The dermal and inhalation routes of entry are likely to be more important to the pesticide applicator than the oral route. You do not purposely eat or drink the chemicals you are using, but you may breathe them in, splash them on your skin, or expose yourself to pesticide fallout. You can be poisoned no matter which way the pesticide enters your body. It may even enter in all three ways and poison you. In other words, if you know the toxicity of a pesticide, you know "how poisonous" it is.

Toxicity of the pesticide chemical is important to human health in terms of the level of exposure; thus, acute toxicity is "how poisonous" one pesticide is, compared to another. It is also the basis for the warning statements on the label. Acute toxicity may be measured as acute oral toxicity, acute dermal toxicity, and acute inhalation toxicity. When properly applied, chlorophenoxy herbicides have presented very minimal hazards to animal life in target areas. The data are expressed as LD_{50} in units of milligrams of chemical per kilogram of body weight. The oral toxicities of 2,4-D and 2,4,5-T are quite similar, e.g., the acute oral LD_{50} of 2,4-D and 2,4,5-T in the rat are 620 and 480 mg/kg, respectively. Tables 1 and 2 summarized the results of several acute toxicity studies with various salt, ester, and amine forms of 2,4-D and 2,4,5-T.

Chronic Toxicity

Chronic toxicity is "how poisonous" a pesticide is to an animal (or man) after small, repeated doses over a period of time. Chronic toxicity is very important because some pesticides can remain in the body for a long time. If you are exposed often to these pesticides, they may remain and build up in your body. You can be poisoned without ever getting a large dose. Chronic toxicity is measured as chronic oral toxicity, chronic dermal toxicity, and chronic inhalation toxicity. Because of the active secretion of chlorophenoxy herbicides, rather large amounts must be administered over a long period of time to produce symptoms of toxicity. In one study,[2] sheep were given 2 g of the acid daily and sacrificed on the day following the final dose. Residues in the tissues were less than 1 ppm in all tissues and usually less than 0.05 ppm which was the sensitivity of the analytical method. Mitchell and co-workers[10] pastured sheep and cattle on treated foliage without harmful effects to the animals. They also fed a lactating cow 5.5 g of 2,4-D daily for 106 days without producing poisoning. Palmer[11] found that cattle were not harmed by 112 daily doses (administered 5 days each week) of 5 mg/kg of alkanolamine salt, but that 44 daily doses of 200 mg/kg or 20 doses of 250 mg/kg would harm the cattle. Table 3 summarizes chronic toxicity studies of phenoxy herbicides.

MEASURING TOXICITY

Finding out the toxicity of a pesticide is not easy. Obviously, humans cannot be test animals. Other animals (usually rats) are used. However, just because a pesticide is

Table 1
ACUTE TOXICITY OF PHENOXY HERBICIDES TO TERRESTRIAL ANIMALS

Derivative	Animal	Dose (mg/kg)	Effect	Ref.
Alkanolamine	Chick	380—765	LD_{50}	1
Isopropyl ester	Rat	700	LD_{50}	1
	Chick	1420	LD_{50}	1
	Guinea pig	550	LD_{50}	1
Butyl ester	Rat	620	LD_{50}	1
	Guinea pig	848	LD_{50}	1
	Chick	2000	LD_{50}	1
PGBE[a]	Rat	570	LD_{50}	1
Acid	Dog	100	LD_{50}	1
	Chick	541	LD_{50}	1
Triethanolamine	Swine	50	No effect	2
	Swine	500	Lethal	2
Butyl ester	Swine	100	No effect	2
Triethanolamine	Chicken	300	No effect	2
Butyl ester	Rat	620	LD_{50}	3
Isopropyl ester	Rat	700	LD_{50}	4
Unspecified amine	Mallard duck	2000	LD_{50}	5
Acid	Pheasant	472	LD_{50}	5
	Mule deer	400—800	LD_{50}	5
	Rat	500	LD_{50}	1
Isopropyl	Mice	551	LD_{50}	1
Butyl ester	Mice	940	LD_{50}	1
Amyl ester	Rat	750	LD_{50}	1

[a] Propylene glycol butyl ether.

very poisonous to rats doesn't mean it is also very poisonous to dogs, cows, man, or wildlife. Likewise, a pesticide which is relatively nontoxic to rats is not always safe to other animals. Toxicity studies are just guidelines. They are used to estimate "how poisonous" a pesticide is compared to another pesticide.

Acute toxicity measure — The commonly used measure of acute oral and dermal toxicity is LD_{50} LD means lethal dose (deadly amount). LD_{50} values are measured from 0 up. The lower the LD_{50}, the more poisonous the pesticide. A pesticide with an acute oral LD_{50} of 500 would be much less toxic than a pesticide with an LD_{50} of 5. Usually LD values are measured in milligrams per kilogram. Often you will note the abbreviation "mg/kg" following the number value. "Milligram" and kilogram" are metric units of weight similar to "ounce" and "ton". Milligrams per kilogram is the same as parts per million. For example, the oral LD_{50} of Pesticide A is 5 mg/kg. Therefore, it would take only 5 parts of Pesticide A for every 1 million parts of body weight to kill a test animal.

Acute inhalation toxicity is measured by LC_{50}. LC_{50} means lethal concentration. "Concentration" is used instead of "dose" because the amount of pesticide inhaled from air is being measured. LC_{50} values are in milligrams per liter. Liters are metric units of volume similar to a quart. The lower the LC_{50} value, the more poisonous the pesticide.

Chronic toxicity measures — There is no standard measure like LD_{50} for chronic toxicity studies. Often the length of experiment in days, months, or years and the amount of each dose is stated. For example, a study of chronic oral toxicity might look this: 8 mg of pesticide were fed to rats daily for 2 years. No symptoms of poisoning appeared. Nonlethal effects of phenoxy herbicides are summarized in Table 4.

Table 2
ACUTE EFFECTS OF 2,4-D DERIVATIVES UPON AQUATIC ANIMALS

Derivative	Animal	Concentration	Effect	Ref.
Isooctyl ester (from 3 manufacturers)	Bluegill	10—31 ppm	48 hr TLm[a]	6
PGBE[b]	Bluegill	17 ppm	48 hr TLm	6
Butoxyethanol ester	Bluegill	1.4 ppm	48 hr TLm	6
PGBE ester	Shrimp	1 ppm (48 hr)	20% mortality or paralysis	7
	Fish (saltwater)	0.32 ppm	48 hr TLm	7
Alkanolamine salt	Bluegill	435—840 ppm	48 hr LC_{50}	8
Dimethylamine salt	Bluegill	166—458 ppm	48 hr LC_{50}	8
Isooctyl ester	Bluegill	8.8—59.7 ppm	48 hr LC_{50}	8
Dimethylamine salt	Fathead minnow	10 ppm	96 hr LC_{50}	8
Acetamide	Fathead minnow	5 ppm	96 hr LC_{50}	8
Oil soluble amine salt	Bluegill, fathead minnow	2 ppm	4 months LC_{10}	8
PGBE ester	Bluegill, fathead minnow	2 ppm	4 months LC_{10}	8
Butyloxyethyl ester	Bluegill, fathead minnow	2 ppm	72 hr LC_{50}	8
Butyl and isoproyl ester, mixed	Bluegill	1.5—1.7 ppm	48 hr LC_{50}	8
N,N-dimethyl cocoamine salt	Bluegill	1.5 ppm	48 hr LC_{50}	8
Ethyl ester	Bluegill	1.4 ppm	48 hr LC_{50}	8
Butyl ester	Bluegill	1.3 ppm	48 hr LC_{50}	8
Isopropyl ester	Bluegill	1.1 ppm	48 hr LC_{50}	8

[a] Median tolerance limit.
[b] Propylene glycol butyl ether.

METABOLISM AND EXCRETION

Most of the data derived from acute toxicity studies indicate that neither 2,4-D nor 2,4,5-T are particularly toxic.[12,15] In the rat, the single-dose, LD_{50} ranges from about 250 to 270 mg/kg, depending on the forms of the chemical administered.[12] Several workers have suggested that part of the reason for this lack of toxicity is that the excretion of the herbicides is very rapid in most mammals.[16,17] Most studies indicate that animals possessing highly developed renal function will rapidly eliminate 2,4-D and 2,4,5-T by active tubular secretion. Cattle and rabbits which normally actively metabolize compounds mostly by acetylation, excrete 2,4,-D and 2,4,5-T in the urine mostly unchanged. Erne[3] found that in the rat, rabbit, calf, and chicken, 2,4-D and 2,4,5-T had a biological half-life varying from 3 to 12 hr and that urinary excretion was the most common route of elimination. Data exist to indicate that only very small amounts of 2,4-D are metabolized by the rabbit.[16,17] Berndt and Koschier[18] studied the in vitro uptake of 2,4-D and 2,4,5-T by the renal cortical tissue of rabbits and rats. Renal cortical sliced from both species accumulates 2,4-D and 2,4,5-T, with greater uptake occurring in rabbit tissue. Nitrogen and various metabolic inhibitors reduced the uptake, thus indicating that both of these organic acid herbicides are transported by the renal organic anion mechanism. Berndt and Koschier[18] concluded that renal tubular transport by the organic anion mechanism may account for the relatively rapid disappearance of these compounds, and this may account for their low toxicity re-

Table 3
CHRONIC TOXICITY OF 2,4-D AND 2,4,5-T DERIVATIVES TO TERRESTRIAL ANIMAL

2,4-D Formulations

Formulation	Animal	Dose	Duration	Effect	Ref.
Triethanolamine	Swine	50 mg/kg/day	3 doses	None	2
	Swine	50 mg/kg/day	8—10 doses	Minor transient effects	2
Butyl ester	Swine	50 mg/kg/day	5 doses	None	2
Triethanolamine	Swine	500 ppm in feed	1 month	Some locomotory disturbance, dedepressed growth rate, no gross pathology	2
	Rat	1000 ppm in water	10 months	Depressed growth rate, no gross pathology	2
	Chicken	1000 ppm in water	Daily from hatching through first 2 months of egg production	Egg size normal, production reduced 30%	2
Alkanolamine	Sheep	100 mg/kg/day	481 days	No effect	9
	Cattle	50 mg/kg/day	112 days	No effect	9
PGBE[a] ester	Sheep	100 mg/kg/day	481 days	No effect	9
Ethylhexyl ester	Cattle	250 mg/kg/day	14 days	Ill in 3 days, survive, and recover from 9 doses; 14 doses lethal	10
	Sheep	250 mg/kg/day	17 days	Ill in 3 days; 17 doses lethal	10
	Sheep and cattle	100 mg/kg/day	10 days	None to minor effects	10
Not specified	Dog	500 ppm in feed	2 years	None	11
	Rat	1250 ppm in feed	2 years	No effects on growth, survival, hermatology, or tumor incidence	11
	Rat	500 ppm in feed	2 years	No effects in reproduction studies	11
Alkanolamine	Chicken	100 mg/kg/day	10 days	No effect on weight gain	12
PGBE ester	Chicken	50 mg/kg/day	10 days	No effect on weight gain	12
	Cattle	100 mg/kg/day	10 days	No effect	12
Acid	Mule deer	80 and 240 mg/kg/day	30 days	Minor symptoms no weight loss	13

2,4,5-T Formulations

Formulation	Animal	Dose	Duration	Effect	Ref.
Not specified	Dog	10 mg/kg/day	5 days/week for 90 days	Minor weight loss, no other effects	14
	Dog	20 mg/kg/day	5 days/week for 90 days	Lethal between 11 and 75 days	14
PGBE ester	Cattle	100 mg/kg/day	10 days	None	12
	Sheep	50 mg/kg/day	10 days	None	12
	Sheep	100 mg/kg/day	369 days	Ill at 367 doses, lethal at 369 (dosed by capsule)	12
PGBE ester	Chicken	100 mg/kg/day	10 days	No effect on weight gain	12
Triethylamine	Sheep	100 mg/kg/day	481 days	No effect	12

[a] Propylene glycol butyl ether.

Table 4
NONLETHAL EFFECTS OF PHENOXY HERBICIDES UPON AQUATIC ANIMALS

Derivative	Animal	Dose	Effect
Butoxyethanol ester	Oyster	3.75 ppm (96 hr)	50% decrease in shell growth
	Shrimp	1 ppm (48 hr)	No effect
	Fish (saltwater)	5 ppm	48 hr TLm[a]
Butoxyethanol	Phytoplankton	1 ppm	16% decrease in CO_2 fixation
Dimethylamine	Oyster	2 ppm (96 hr)	No effect on shell growth
	Fish	15 ppm	No effect
	Phytoplankton	1 ppm	No effect on Co_2 fixation
Ethylhexyl ester	Oyster	5 ppm (96 hr)	38% decrease in shell growth
	Shrimp	2 ppm (48 hr)	10% mortality or paralysis
	Fish (saltwater)	10 ppm (48 hr)	No effect
	Phytoplankton	1 ppm (4 hr)	49% decrease in CO_2 fixation
PGBE[b] ester	Oyster	1 ppm (96 hr)	39% decrease in shell growth
	Shrimp	1 ppm (48 hr)	No effect
	Fish (saltwater)	4.5 ppm	48 hr TLm

[a] Median tolerance limit.
[b] Propylene glycol butyl ether.

Extracted from Gleason, M. N., Gosselin, R. E., Hodge, H. C., and Smith, R. P., Eds., *Clinical Toxicology of Commercial Products,* 3rd ed., Williams & Wilkins, Baltimore, 1969.

quired to produce fatal poisoning. Palmer and Radeleff[13] reported that sheep were given 481 daily doses of 100 mg/kg 2,4-D without producing poisoning. 2,4,5-T has not been investigated as thoroughly as 2,4-D, but the reaction of cattle and sheep to massive doses would indicate that absorption and excretion must follow a similar pattern.

Gehring et al.[19] studied the effects of 2,4,5-T at a dose level of 5 mg/kg ingested directly or in a slurry of milk. Analytical-grade 2,4,5-T, having a purity of greater than 99% and containing less than the detectable level 0.05 ppm of TCDD, was used. Complete medical histories and physical and laboratory studies were accomplished before and repeated after the study. It was found that the clearances of 2,4,5-T and the excretion from the body were by first-order rate processes with half-lives of 23.10 and 23.06 hr respectively. Essentially all of the ingested 2,4,5-T was absorbed into the body and was excreted unchanged in the urine. Following ingestion, 65% of the 2,4,5-T remained in the plasma, where 98% was reversibly bound to the plasma proteins.[19] No untoward effects associated with the ingestion of 5 mg/kg 2,4,5-T were detected in any of the subjects. A metallic taste lasting 1 to 2 hr following ingestion was reported by most of the subjects. It was also concluded that essentially all of the ingested 2,4,5-T was absorbed and then eliminated unchanged in the urine.

The behavior of the chloropenoxy herbicides in nonmammalian aquatic animals is quite different than the behavior described for terrestrial mammals and birds. The herbicides have a greater toxic potential for aquatic animals. First, the route of entry is different in most instances. The aquatic animal absorbs the herbicide which is distributed throughout his total environment (absorption is mainly via gills in fish). Then, the differences in renal function must be considered. Generally, nonmammalian aquatic animals do not have highly developed kidneys. Thus, once the herbicide is in the aquatic animal's body, some metabolic changes must occur in the molecule to make it more polar if it is to be excreted. Toxicity testing is also necessarily different with aquatic animals. Usually, aquatic animals are placed in a concentration of the toxicant to gradually absorb the material at a rate depending on the animal's physiology and the behavior of the toxicant in the particular water conditions. Therefore the actual dose to each animal is not known in most studies with aquatic animals. In contrast, toxicity studies with terrestrial animals usually allow calculation of a known dose per unit weight of each animal. Thus, toxicities are often reported as "LD_{50}" for terrestrial animals and "LC_{50}" for aquatic animals.

Schultz[20] studied the uptake, distribution, and dissipation of C-labeled dimethyl amine salt of 2,4,-D (DMA-2,4-D). Three species of fish were exposed to a 0.5, 1,0, or 2.0-mg/ℓ concentration of herbicide for an exposure period of 84 days. No mortalities occurred, nor were adverse biological effects observed at these exposure levels. The highest radioactive residue found in muscle tissue occurred in bluegills exposed to 2.0 mg/ℓ for 84 days (1.065 mg/kg). However, gas-liquid chromatography indicated that over 90% of radioactive residues consisted of metabolites of 2,4-D. The major metabolite in the fish was found to be 2,4-D glucuronic acid conjugate. It was postulated that the herbicides excreted by the fish were degraded by microflora of the growth pool and resorbed as radioactive carbon dioxide fragments. Sikka et al.[21] demonstrated that this was indeed the case. The growth pool studies by Schultz were repeated by Stalling and Huckins.[22] They demonstrated that the radioactive carbon dioxide fragments were incorporated into the fatty acids of the fish flesh and that these residues are not radioactive herbicide as such.

SOLUBILITY AND HYDROLYSIS

The esters of 2,4-D or 2,4,5-T have a very limited solubility in water. Because of this very low solubility, the actual concentrations of esters produced in a body of water by accidental contamination would likely be very low. In one study using artificial seawater, the herbicide was mixed into the water in an amount equal to 150 mg/ℓ. Had all components gone into solution, ester concentrations would have been 64 (2,4,-D NBE) and 61 mg/ℓ (2,4,5-T NBE). The actual measured concentrations were 2 (2,4-D NBE) and 1.8 mg/ℓ (2,4,5-T NBE) immediately after mixing. These increased to 18 and 22 mg/ℓ of 2,4-D NBE and 2,4,5-T NBE, respectively, at 24 hr and then started a rapid decline to 7.5 and 9.5 mg/ℓ at 48 hr after mixing. The rate of disappearance of the ester of 2,4-D was fairly rapid and was assumed to be mainly a result of hydrolysis. The half-life of the ester was 15 hr.

It is important that when the esters of 2,4-D and 2,4,5-T hydrolyze, their toxicity to aquatic animals is decreased by almost a factor of 10. In the static situation, the rate of hydrolysis is probably faster than the rate that the ester went into solution so that lethal concentrations were never attained. Toxicity studies with freshwater and saltwater animals have been the so-called "static bioassay" in which no attempt is made to maintain a constant concentration of the herbicide ester in each test chamber. "Concentrations" are theoretical and based on volumes of herbicide and water mixed together rather than from analysis of water to quantitate the herbicide. Most studies reported from literature are of the same type. The toxicity tests revealed that in both freshwater and saltwater, most of the test organisms had responded at 12 hr of exposure. There was rarely any increase in mortality past 24 hr.

Many other factors can influence the concentration of *N*-butyl esters of 2,4-D and 2,4,5-T in a body of water. In studies where large amounts of Orange herbicide were placed in water, the globules of the herbicide appeared to become coated with an opaque material that may have inhibited the ester from going into solution.

METABOLISM IN PLANTS

2,4-D and 2,4,5-T are systematic herbicides that affect plants by hormonal type of action usually described as "auxin-like" or "auxin-type". These chemicals are a group of substances which promote plant growth by cell elongation, bring about root formation, or cause bud inhibition or similar effects. When applied to leaves of a plant, chlorophenoxy herbicides are absorbed through the cuticle into the plant system.

The *N*-butyl ester forms of 2,4-D and 2,4,5-T are usually more effective than more polar forms because of better absorption into the plant. Approximately 10 times as much 2,4-D was absorbed from a medium having pH 3 than one with pH 11. 2,4-D has a pK(a) of 2.8 and would be highly disassociated at pH 11. Once the herbicide is in the plant, it is translocated to areas where food is being stored as in rapidly growing new roots and shoots. The chlorophenoxy herbicides can be stored in certain cells of the plant. Also, metabolism occurs through degradation of the acetic acid side chain, hydroxylation of the aromatic ring, or conjugation.

Once in the plant, herbicides act by interfering with the photosynthetic, respiratory, and other plant processes causing the plant to lose its leaves and ultimately die. Plant susceptibility to sublethal exposures of 2,4-D is markedly influenced by the growth condition of the plant and by environmental factors. Since most of the injury is expressed by growth response, the plant must be growing in order to show injury. In addition, plants in shaded areas respond more slowly than those exposed to direct sunlight. Because of these various factors which affect plant response to the 2,4-D-type herbicide, differences in lists showing plant susceptibility should be expected. Orange herbicide is effective on a wide variety of woody and broadleaf plant species. Other lower plant forms can also be affected by auxin-type herbicides. However, much higher doses of herbicides are required than for plants with a more complex structure.

TOXICOLOGICAL TESTING

Insistence on administering a "maximum tolerated dose" may be very misleading if this is the only dose tested.

There is no justification for abrogating the need to establish a dose-response relationship which is fundamental to all toxicological experimentation. The route of administration is all important in tests for teratogenesis. We are told that "peritoneal administration is an appropriate test route for pesticides to which humans are exposed by inhalation, or for pesticides which are systemically absorbed, following ingestion". It is safe to predict that, by appropriate choice of dose, concentration of solution, and frequency of administration by subcutaneous (s.c.) route, any chemical agent can be shown to be a carcinogen or a teratogen in the rat and probably in other laboratory rodents.

The bionetics study began with the observation that 2,4,5-T was teratogenic and feticidal in two strains of mice when administered either s.c. or orally and was teratogenic and feticidal in one strain of rats when administered orally.[24] Analyses of the sample of 2,4,5-T that had been tested against the animals revealed the presence of 27 ± 8 ppm tetrachlorodibenzo-*p*-dioxin (TCDD). Subsequent study of standard 2,4,5-T containing less than 1 ppm TCDD given to rats by gavage in doses up to 24 mg/kg daily failed to reveal evidence of teratogenic or embryotoxic effect.[25] Under similar conditions, TCDD produced no effect at a dose of 0.03 g/kg/day, while doses of 0.125 g/kg/day or greater manifested toxicity to the fetus, and at 8.0 g/kg/day toxicity was manifested to the mother also.[26]

The metabolism of test compound is a highly relevant consideration in teratogenesis. If the metabolic pathway in the test animal differs radically from that in man, then the results of study are unlikely to be useful for assessment of hazards arising from trace contaminants. The finding of teratogenesis or embryotoxicity has meaning only in the appropriate animal species.[23]

ENVIRONMENTAL CONTAMINATION

The possibility that an extraordinarily toxic contaminant of a widely used herbicide may be sufficiently stable in the environment and soluble in fat or other tissues to enter food chains and ultimately the human diet is worthy of consideration. It was known, of course, that 2,4,5-T does not accumulate to any significant degree in animal tissues. Chlorinated dibenzo-*p*-dioxins long have been recognized as by-products from the manufacture of certain chlorinated phenols. For example, 2,4,5-trichlorophenol is prepared industrially by the hydrolysis of 1,2,4,5-tetrachlorobenzene at elevated temperatures and pressures, a process which can also result in the formation of traces of heterocyclic impurities including 2,3,7,8-TCDD if temperatures are permitted to exceed 160°C and if the reaction becomes alkaline. This dioxin is toxic, teratogenic, and acnegenic, and its presence appears to account satisfactorily for the alleged teratogenic effects of trichlorophenol derivatives such as the herbicide 2,4,5-T.

No proven instance of toxicity associated with 2,4,5-T intake in man has been found in agricultural or industrial workers known to have had repeated, relatively high levels of exposure to 2,4,5-T of low dioxin content. The safety factor for the general population is estimated to be several orders of magnitude greater than that for 2,4,5-T factory workers. Data are too limited for a firm conclusion, but there is no evidence to suggest that TCDD as a contaminant in 2,4,5-T is likely to be encountered by animal or man in sufficient dosage to cause toxic reactions.[27]

Since most chemicals under suitable laboratory conditions could probably be demonstrated to have teratogenic effects, and certainly all could be shown to produce some toxic effects if dosage were raised high enough, it would not be reasonable to consider the demonstration of toxic effects under conditions of greatly elevated dosage sufficient grounds for prohibiting further use of particular chemical.[23]

The question of the formation of TCDD as a result of the pyrolysis or burning of wood, including brush treated with 2,4,5-T, has been a matter of some concern. Langer[28] states, "The derivatives of 2,4-D, 2,4,5-T and Silvex as well as their sodium salts and esters have not produced dioxins in pyrolytic reactions whether carried out in the solid state, in the metal, or in solution. Even after conditions of extreme hydrolysis, followed by pyrolysis, we could observe only trace amounts of dioxins." Langer[28] further stated, "even extreme conditions such as burning of treated wood or vegetation after the use of 2,4-D, 2,4,5-T, or silvex is not expected to produce detectable amounts of dioxins or dibenzofuran.

LABEL WARNING STATEMENTS

To alert pesticide users to the toxicity of pesticide, a signal word must appear on most labels. Four different categories are used. Signal words are used to tell the user whether the chemical is highly toxic, moderately (medium) toxic, slightly toxic, or relatively nontoxic. Label warning statements are based on acute toxicity and are all used to determine the correct category of each pesticide. For example, the acute oral toxicity and acute dermal toxicity of a pesticide are in the slightly toxic category. However, the acute inhalation toxicity is in the highly toxic category. The pesticide label will have the signal words for a highly toxic pesticide.

Hazard of Use

This is the chance that danger or harm will come to the applicator, by-standers, consumers, livestock, wildlife, or crops, etc. from the use of pesticide. Hazard is often confused with toxicity. They are not the same. A highly toxic pesticide is usually considered "hazardous" because of the risk that the applicator, by-standers, etc. can be injured. However, safe and careful handling of a highly toxic pesticide can make it a low risk or low hazard, Many factors besides toxicity can make a pesticide hazardous. These include the target involved, the formulation chosen, the concentration and dosage used, the type of pesticide, and the skill of the applicator.

Industrial Exposure

Dow Chemical Midland, Mich. prepared an extensive health inventory of 126 manufacturing personnel in an effort to identify harmful effects of inhaled 2,4,5-T. The inhalation rate of the agent was estimated to be from 1.6 to 8.1 mg/day per worker, depending on work assignment, for periods of up to 3 years. The survey indicates that no illness was associated with 2,4,5-T intake. In plants where 2,4,5-T contained a high proportion of TCDD, Bleiberg et al.[29] found 18% of the exposed employees suffered from moderate to severe chloracne, the intensity of which correlated significantly with the presence of hyperpigmentation, hirsutism, and eye irritation. In the late 1940s, a pressure overload resulted in the accidental rupture of a vessel containing the sodium salt of 2,4,5-trichlorophenol, a precursor of 2,4,5-T. During the following months, 228 persons developed chloracne, not only plant employees, but members of their families, including wives and children. In workers more intensively exposed as a result of the accident, chloracne appeared in about 2 weeks followed by moderate to severe pain in the skeletal muscles of the legs, arms, back and breast, decreased libido, and intolerance to cold. Comedones appeared in areas of adult hair which is not typical of juvenile acne. There were pustules on the face, neck, abdomen, back, and scrotum. Serum lipids, prothrombin time, and glucuronates were all elevated. Biopsy of peripheral nerves revealed destruction of myelin sheaths and in some instances nerve fibers. Hyperpigmentation, fatigue, and marked nervous irritability appeared. Over a period of several months, all of the symptoms and findings, except the scars of acne, returned to normal after removal from exposure. Cases in the families of the workers probably resulted from contaminated clothing and poor personal hygiene.

Target Species

A pesticide is hazardous if it is used on a plant or animal which can be easily injured by it. If a pesticide does not injure plants, there is a low hazard when using the chemical on or around them. However, that same chemical could be highly toxic to animals. If used on or around animals, there is a bigger risk than when used on or around plants. There is evidence that some 2,4-D compounds may be present in the ambient atmosphere in some parts of U.S. at levels sufficient to cause adverse growth effects on sensitive vegetation. During 1962 through 1964, 2,4-D concentrations were measured[12] in air samples taken in an agricultural area of eastern Oregon. Concurrently, they surveyed for auxin-like plant damage in the areas where the air samples were taken.

In the spring of 1962, measured concentrations of the isopropyl ester of 2,4-D in the air ranged from 0.015 to 0.64 ppm. This was during the time of year when the huge wheat fields of the area were being treated for weeds by aerial application of the isopropyl ester. Plant damage to tomato crops appeared to coincide with periods of highest measured concentrations of the isopropyl ester. Other plants, especially locust trees, also showed growth regulator symptoms.

SUMMARY AND CONCLUSIONS

The use of herbicides has been proven a productive means of controlling plants. However, herbicides are not without toxic properties to man. Herbicides have three different effects on man, depending upon whether the point of contact is through the mouth, nose, or skin.

There are two means of measuring the level of toxin, acute toxicity, and chronic toxicity. Usually acute toxicity is measured by the LD_{50} value, while chronic toxicity is measured by the LC_{50} value. The level of toxicity in animals depends upon how developed the metabolism and elimination system is. Those with the higher developed system will either break down or eliminate the toxin faster than those who are less developed.

When examining the different effects of herbicides, one must take into consideration the possibility of chemical reactions involved when the herbicide comes into contact with the environment. For example, a herbicide will hydrolize in water, while another type of reaction will occur on land.

The herbicide actually affects a plant by interfering with photosynthetic, respiratory, and other metabolic processes. This causes loss of the leaves and eventual destruction of the plant.

When testing herbicides, it is important to realize that different amounts of the same herbicide produce varying degrees of toxicity to the plant, as well as to the environment.

There are four signal words used on labels of herbicides to describe the level of toxicity:

1. Highly toxic
2. Moderately toxic
3. Slightly toxic
4. Relatively nontoxic

There are also many factors other than toxicity which make a herbicide hazardous. They include

1. The target involved
2. The formulation chosen
3. The concentration and dosage used
4. The type of pesticide
5. The skill of the applicator

When using herbicides, it is important to know it's entire scope in order not to harm the environment.

The effects of herbicides 2,4-D and 2,4,5-T on man have been minimal, even in large doses. When the various adverse effects of man do appear, they are usually seen in the nervous system or the skin. The symptoms appear shortly after exposure and usually disappear within a few months. It is safe to conclude that the use of herbicides is a productive means of controlling plants with minimal adverse effects to man.

REFERENCES

1. **Rowe, V. K. and Hymas, T. A.**, Summary of toxicological information on 2,4-D and 2,4,5-T type herbicides and an evaluation of the hazards to livestock associated with their use, *Am. J. Vet. Res.*, 15, 622, 1954.
2. **Bjorklund, N. E. and Erne, K.**, Toxicological studies of phenoxyacetic herbicides in animals, *Acta Vet. Scand.*, 7, 364, 1966.
3. **Erne, K.**, Distribution and elimination of chlorinated phenoxyacetic acids in animals, *Acta Vet. Scand.*, 7, 240, 1966.
4. **Erne, K.**, Animal metabolism of phenoxyacetic herbicides, *Acta Vet. Scand.*, 7, 264, 1966.
5. **Tucker, R. K. and Hudson, R. H.**, 2,4,5-T Acute Oral Toxicity to Mallard Ducks, Denver Wildlife Research Center, U.S. Department of the Interior, Colorado, 1970.
6. **Hughes, J. S. and Davis, J. T.**, Variations in toxicity to bluegill sunfish of phenoxy herbicides, *Weeds*, 11, 50, 1963.
7. **House, W. B., Goodson, L. M., Gadberry, M. H. and Docter, K. W.**, Assessment of Ecological Effects of Extensive or Repeated Use of Herbicides: Final Report, ADX24314, Midwest Research Institute, Kansas City, 1967.
8. **Lawrence, J. M.**, Aquatic Herbicide Data, Agriculture Handbook No. 231, U.S. Department of Agriculture, Agricultural Research Service, Agriculture Experiment Station, Auburn University, Ala., 1964.
9. **Palmer, J. S., and Radeleff, R. D.**, The toxicologic effects of certain fungicides and herbicides on sheep and cattle, *Ann. N.Y. Acad. Sci.*, 111, 729, 1964.
10. **Mitchell, J. W., Hodgson, R. E., and Gaetjens, C. F.**, Tolerace of farm animals to feed contaminated by 2,4-D, *J. Anim. Sci.*, 5, 226, 1946.
11. **Palmer, J. S.**, Chronic toxicity of 2,4-D alkanolamine salts to cattle, *J. Am. Vet Med. Assoc.*, 143, 398, 1963.
12. **Pimentel, D.**, Ecological Effects of Pesticides on Non-Target Species, U.S. Government Printing Office, Washington, D.C., 1971.
13. **Palmer, J. S. and Radeleff, R. D.**, The Toxicity of Some Organic Herbicides to Cattle, Sheep, and Chickens, Product Res. Rep. No. 106, Agricultural Research Service, U.S. Department of Agriculture, Washington, D.C., 1969.
14. **Palmer, J. S.**, Toxicity of 45 Organic Herbicides to Cattle, Sheep, and Chickens, Product Res. Rep. No. 137, U.S. Department of Agriculture, Agriculture Research Service, Washington, D.C., 1972.
15. **Gleason, M. N., Gosselin, R. E., and Hody, H. C.**, *Clinical Toxicology of Commercial Products*, 3rd ed., Williams & Wilkins, Baltimore, 1969.
16. **Christensen, H. E., Ed.**, Toxic Substances, U.S. Government Printing Office, Washington, D.C., 1971.
17. **Clark, D. E., Young, J. E., Younger, R. L., Hunt, L. M., and McLaren, J. K.**, The fate of 2,4-dichlorophenoxyacetic acid in sheep, *J. Agric. Food Chem.*, 12, 43, 1964.
18. **Berndt, W. O. and Koschier, F.**, In vitro uptake of 2,4-dichlorophenoxyacetic acid (2,4-D and 2,4,5-trichlorophenoxyacetic acid (2,4,5-T) by renal cortical tissue and rabbits and rats, *Toxicol. Appl. Pharmacol.*, 26, 559, 1973.
19. **Gehring, P. J., Kramer, C. G., Schwetz, B. A., Rose, J. Q., and Rowe, J. K.**, The fate of 2,4,5-trichlorophenoxyacetic acid (2,4,5-T) following oral administration, *Toxicol. Appl. Pharmacol.*, 26, 352, 1973.
20. **Schultz, D. P.**, Dynamics of a salt of (2,4-dichlorophenoxy) acetic acid in fish, water and hydrosol, *J. Agric. Food Chem.*, 21, 186, 1973.
21. **Sikka, H. C., Appleton, H. T., and Gangstad, E. O.**, Uptake and metabolism of dimethylaminesalt of 2,4-dichlorophenoxyacitic acid by fish, *J. Agric. Food Chem.*, 25, 1030, 1977.
22. **Stallings, L. D. and Huckins, J. N.**, Metabolism of 2,4-dichlorophenoxyacity acid (2,4-D) in bluegills and water, *J. Agric. Food Chem.*, 26, 447, 1978.
23. **Goldberg, L.**, Trace chemical contaminants in food: potential for harm, *Food Cosmet. Toxicol.*, 9, 65, 1971.
24. **Courtney, K. D. and Moore, J. A.**, Teratology studies with 2,4,5-trichlorophenoxyacetic acid and 2,3,7,8-tetrachlorodibenzo-*p*-dioxin, *Toxicol. Appl. Parmacol.*, 20, 396, 1971.
25. **Emerson, J. L., Thompson, D. J., Strebling, R. J., Gerbig, C. G., and Robinson, J. B.**, Teratogenic study of 2,4,5-trichloro-phenoxyacetic acid in the rat and rabbit, *Food Cosmet. Toxicol.*, 9, 395, 1971.
26. **Sparschu, G. L., Dunn, F. L., and Rowe, V. K.**, Teratogenic study of 2,3,7,8,-tretrachlorodibenzo-*p*-dioxin in the rat, *Toxicol. Appl. Pharmacol.*, 17, 317, 1970.

27. Advisory Committee on 2,4,5-T, Report of the Advisory Committee on 2,4,5-T to the Administrator of the Environmental Protection Agency, U.S. Environmental Protection Agency, Washington, D.C., 1971.
28. Langer, H. G., The information of dibenzodioxins and other condensation products from chlorinated phenols and derivatives, *Environ. Health Perspect.*, 5, 3, 1973.
29. Bleiberg, J. M. Wallen, M., Brodkin, R., and Applebaum, I. L., Industrially acquired porphyria, *Arch. Dermatol.*, 89, 793, 1964.
30. Turner, D. J., The Safety of the Herbicides 2,4-D and 2,4,5-T, Agricultural Research Council Weed Research Organization, London, 1979.

Chapter 15

ECOLOGICAL SIGNIFICANCE OF RIGHTS-OF-WAY WEED CONTROL

INTRODUCTION

The rights-of-way for power, communication, roadways, and pipelines are unusual noncropland areas in that they are a continuous network extending for thousands of miles over the U.S. The rights of way pass through all of the environments represented by a broad spectrum of climatic and edaphic systems.

The vegetation along the rights-of-way is as diverse as all the vegetation patterns represented in the country, and the land may exist relatively undisturbed by man for long periods of time. In contrast, roads and railroads represent an entirely different ecological system which is continuously disturbed by many different factors.

The greatest change that takes place in rights-of-way from an ecological viewpoint is the initial disturbance of the vegetation. The disturbance of the fauna and lower forms of organisms is secondary, but there are complex interactions between trees, shrubs, herbs, animal life, arthropods, mollusks, fungi, and other organisms. The ecological relationships among these biota and the abiotic factors of the ecological relationships among these biota represent an ecosystem of many different types of subclimax vegetation.

Power, communication, and pipeline rights-of-way are usually cleared initially by mechanical means, whereas thereafter they are maintained by herbicidal treatment, combined in some instances with mechanical methods. The maintenance of rights-of-way is necessary along power and communication lines in order to keep the area close to the wires free of vegetation. If foliage is allowed to return to its original state, access would be denied maintenance crews and there would be a risk of "shorting" occurring between high-voltage transmission lines and the vegetation.

Central part of right-of-way— Directly under the wires, all tall-growing trees which might eventually grow or fall into the lines are removed. Low-growing shrubs and scattered taller shrubs are preserved. In general, shrubs that eventually grow over 3 ft in height should be removed; they occur in large colonies which spread over the ground for distances of 25 ft or more. A trail 8 to 10 ft in width, preferably composed of grasses, should be maintained in order to permit access for inspection and repair. Similar pattern is recommended around the bases of poles and towers.

The sides of the right-of-way— Between the outermost wires the forest edge, tall-growing trees which might eventually grow or fall into the lines are removed. Therefore, along the sides all shrubs and low-growing trees are preserved to form the densest possible cover.

Forest edge adjacent to the right-of-way— Here tall-growing trees are removed before they reach the height of the wires. This eliminates the danger of trees falling into the wires and avoids costly cutting operations in the future.

ECOLOGICAL INTERESTS AND VALUES

There are a number of interests involved in maintenance of rights of way

1. The utility company
2. The landowner, who has granted the easement
3. The county and state highway commission

4. General public, who have a high exposure to rights-of-way traffic

There is probably no form of vegetation management that is more obvious to the general public than that exercised along our vehicular rights of way. The conservation-minded public and the land owner, and in some sense the general public, are interested in these areas from the standpoint of the values that are involved in these areas, if properly managed.

General conservation values— These values are recognized by all conservation-minded groups. They are important for the greatest social good — now and in the future. These values have had great public response in the last decade.

Wildlife habitat— This is the concern of biologists in general, involving various clubs and societies. Here, wildlife includes song birds. The wildlife habit has received general public attention.

Game habitat— This is the chief interest of the various state game departments as well as sportsmen and sportsmen's clubs. The values of the rights-of-way are for the production of and to increase game populations. Sportsmen are keenly aware of the significance of this edge effect in promoting game development.

Forestry values— The concern of foresters is primarily in obtaining suitable stands adjacent to the rights-of-way and removing threats of disease and insects to timber species. Access to the timberland is from the rights-of-way approaches.

Habitat for flowering herbs and rare plants— This is of particular interest to state botanical societies and professional and amateur naturalists and botanists. The preservation of species or associations that might otherwise become extinct is of primary importance.

Landscape and aesthetic values— This is the concern of garden clubs and the general public, especially where lines cross or are seen from public roads.

CONTROL TECHNIQUES

There have been three relatively long-term controlled vegetation and ecological studies carried out on definitive areas of vegetation: Aton forests (400 acre) in northwestern Connecticut initiated in 1946; the Connecticut Arboretum right-of-way demonstration initiated in 1953; and a 3-mi section of the Pennsylvania Electric Company's power line in Center County, started in 1953. The objective of this work was to develop a vegetation-control management system that would lead to a stabilized plant community. The approach was to selectively spray to promote the growth of a varied mixture of grasses, forbs, and shrubs. The development of shrublands was considered by Egler to be closed communities not being invaded by tree seedlings. He felt that the summer forage blanket spraying was not successful in controlling the growth from root-suckering species, such as black locust, quaking aspen, and sassafrass.[1-7]

Some principles of vegetation management indicate that rights of way are nothing more than nonforested land which tend to develop into forests. All new disturbed areas tend to progress more or less quickly to some forest type. The normal course of development from open land to forests passes through the following stages:

1. Annual weeds
2. Grasslands
3. Shrublands
4. Forests

There were two interpretations of physiognomic development, namely "relay floristics" and "initial floristics" composition. The theory of the former assumes that the invasion of new species is in turn killed out by the species of the previous stages, assuming that the sequential invaders, weeds, grasses, forbs, shrubs, and trees, are all present on or in the soil at the beginning following an overgrazing, a fire, or other types of destruction, existing in the form of seeds, seedlings, or shoot-producing roots.

In brush field reclamation, the ability of plant species to germinate, survive, grow, and reproduce in a habitat depends upon their efficiency to compete for light, soil moisture, and nutrients. Competition for light is the function of the above-growing portions of the plant, whereas the competition for soil moisture and mineral nutrients are confined to the root systems. The competition for nutrients by root systems is not as severe as competition for soil moisture.

The techniques used at the Connecticut Arboretum for managing a utility right-of-way include knapsack sprayers as well as commercial power equipment. Basal treatments give effective root kill in all seasons, but the root collar applications are not effective. White oak showed some resistance to root kill. Stem foliar treatments are not too successful against sassafras and aspen, and they frequently resurged profusely after spraying. Late or midsummer basal or root collar treatments are effective on aspen as well as sumac. Low shrubs, such as huckleberry, low bush blueberry, sweet fern, and others, which have a tendency to form a tight ground cover and prevent the invasion of tree seedlings are preserved. Also the shrubs and low trees that are left provide excellent food and cover for wildlife.

The technique used for spraying of power and communication right-of-way is the use of helicopters, fixed-wing aircraft, and mobile units along the roads. It is common practice in the southern deciduous pine forests to spray the rights-of-way every 3 to 5 years. The spraying is controlled to reduce the height of the vegetation, and much of the right of way does not revert to a grassland stage. The advent of aerial application has solved the difficult logistics of transporting herbicides by carriers for long distances into inaccessible areas. In recent times, there are two problems involving the cost of right-of-way maintenance that have become more paramount than all other factors, and that is cost of and the lack of sources of qualified manual labor.[6,7]

PROGRAMS AND COST

Agricultural practice in recent times has been characterized by the elimination of the smaller farms in order that the utilization of modern high-priced farm equipment might be more efficient. The farm hand, the "hired hand" of the 1920 era and before, has almost totally disappeared from the scene. The same considerations apply to the labor force for right-of-way control.

There is a question whether the selective basal or root collar technique is feasible from an economic standpoint. This approach to vegetation management appears to have been successful at least over the period of time that reports are available. It would be valuable for its complete appraisal if cost data were available.

Some comments concerning the use of knapsack mistblowers on the National Lead Company main power transmission line rights-of-way in northern New York is of interest here (1959 to 1965). The selection of an optimum approach to brush control for set of circumstances involves the skillful blending of two factors: treatment cost and long-term effectiveness. A very expensive treatment, such as cutting and stump spraying, may be justified if it provides adequate control for sufficient long period of time.

They used a low-volatile ester of 2,4,5-trichlorophenoxy (2,4,5-T), containing 4 lb of acid per gallon. They sprayed the 100 ft right of way, utilizing three operators working abreast. Daily ground coverage of three operators ranged from 10 to 25 acres. The cost breakdown for herbicides, labor, supervision, vehicles, and mistblower expense for a 200-acre plot was $27.48/acre in 1967.

Problems concerning mistblower operation are

1. Application of 2,4,5-T left softwood stems largely unscathed.
2. Limiting problem in small areas with tall brush, up to 30 or more feet in height; they resorted to using ultralight direct-drive chain saws.
3. A total of 20% of the cost of brush control was distributing the bush chemical along the rights-of-way.

In some studies on ecological aspects, the objectives were

1. To determine the effectiveness of chemical brush control on game food and cover
2. To stabilize the game usage of the treatment areas
3. To follow the effectiveness the original treatment of brush control with and without subsequent sprays

In 1 year following the initial spray, a follow-up basal spray was applied to one half of each treatment area of five original treatments. Treatments and follow-up sprays were as follows:

1. Unsprayed, cut as needed for control
2. Broadcast foliage spray of 2,4-dichlorophenoxy (2,4-D) plus 2,4,5-T butoxyethanol esters, half and half, at a concentration of 4 lb [acid equivalent (ae) per 100 gal] in water (applied in June)
3. Oil-water, semibasal spray of emulsifiable acids of 2,4-D plus 2,4,5-T, half and half, 3 gal of spray material to make a concentration of 6 lb ae/ha in an oil-water carrier consisting of 10 gal of Number 2 fuel oil in 87 gal of water (applied in June)
4. General summer basal spray of emulsifiable acids of 2,4-D plus 2,4,5-T, half and half, at a concentration of 12 lb ae/ha in Number 2 fuel oil (applied in June)
5. Selective winter basal spray of 2,4,5-T butoxyethanol esters at a concentration of 12 lb ae/ha in Number 2 fuel oil (applied in February)
6. Broadcast foliage spray of ammate at a concentration of 3/4 per gal of water; 4 oz of DuPont® sticker-spreader added per 100 gal of spray (applied in June)

In regard to the control of woody plants of all the spray treatments listed above, the semibasal spray was superior to the other treatments.

The summer and the winter basals were the poorest due to the lack of effectiveness of controlling sassafras. A combination of an initial spray plus a follow-up basal spray was more effective than the initial spray. The best control was obtained in the 2,4-D + 2,4,5-T plus basal spray plots. The semibasal and winter basal with a follow-up basal spray were only slightly poorer than the other treatments.

One of the objectives of the experiment was to develop a stable ground cover. The selective basal sprays were highly successful in maintaining the original bracken-hedge-blueberry plant community. Broadcast spray areas returned to essentially the original plant components in 5 years.

Trends in animal populations in the various test plots, highly important from an ecological viewpoint, are summarized as follows:

1. Deer — observed in all areas and the total usuage of these areas increased
2. Grouse — observed in all areas
3. Turkeys — used treatment areas, broadcast foliage spray areas, flocks observed in the summer when they fed upon insects

4. Squirrels — use of rights-of-way along the edge where most producing trees were concentrated
5. Rabbits — increased in the treatment areas

All the major species of plants composing the dominant plant were used by the major game species found in the area.[8-16]

SUMMARY AND CONCLUSIONS

Major nonagricultural uses of herbicides include the management of vegetation on roadsides, rights-of-way, transmission lines, fire breaks, and parking areas. Brush, trees, and other vegetation may interfere with transmission lines, block views, hinder passage along roadways, or restrict other activities on rights-of-way. The largest industrial use of herbicides is for brush control under transmission lines.

Brush and young trees can be effectively controlled by herbicides or mechanical means alone. When the concept of management is substituted for the concept of control, prescriptions can be written to achieve the desired effect, and these prescriptions can include mechanical methods, herbicides, or both. In extremely rough terrain where it is impossible to use tractors and other heavy mechanical control equipment, herbicides are almost a necessity. On flat terrain, if only grass cover is desired, mechanical means may be equally effective. Depending on the terrain and availability of labor in a region, weed control is generally less expensive with herbicides than with mechanical means.

In the management of the rights-of-way, repeated blanket spraying produces and maintains low ground cover. Selective basal spraying, however, produces a habitat of maximum diversity, maintaining a variety of shrubs and herbs that produce fruit and seeds and attract varied animal life. Retreatments are needed much less frequently than with blanket spraying, and costs are therefore essentially equal over a period of years. Herbicides may pollute environments adjacent to treated rights-of-way by drift during application and by volatilization or runoff after application; the principal problem is drift. Because rights-of-way are normally a few hundred feet wide and extend for many miles through wild habitats, the potential for polluting adjoining habitats is great.

The elimination of some vegetation with herbicides may expose portions of the soil to runoff and soil erosion. Since many grasses cover the bare areas quickly, this condition lasts only a part of a season. With primarily a grass cover, the water yield measured in stream flow from the treated watershed areas may be increased as much as 80%. Although mechanical control may result in reduced runoff in some cases, running tractors over the soil may leave wheel tracks that collect water and eventually result in small washed gullies. Management practices for woody-plant control that bare the soil in rights-of-way are detrimental in the long run. Through the judicious use of selective herbicides, total ecosystems can be maintained.

REFERENCES

1. **Egler, F. E.**, Effects of 2,4-D on woody plants in Connecticut, *J. For.*, 45, 449, 1947.
2. **Egler, F. E.**, 2,4-D Effects in Connecticut vegetation, *Ecology*, 29 (3), 382, 1948.
3. **Egler, F. E.**, Herbicide effects in Connecticut vegetation, 1949, *Bot. Gaz. (Chicago)*, 112, 76, 1950.
4. **Niering, W. A.**, Chemical control of woody species: a summary, *Proc. Northeast., Weed Control Conf.*, 10, 212, 1956.
5. **Niering, W. A.**, The Connecticut Arboretum right-of-way demonstration area progress report, *Proc. Northeast. Weed Control Conf.*, 11, 203, 1957.

6. **Niering, W. A.,** Principles of sound right-of-way vegetation management, *Econ. Bot.,* 12, 140, 1958.
7. **Niering, W. A.,** The Connecticut Arboretum right-of-way demonstration area — its role in commercial application, *Proc. Northeast. Weed Control Conf.,* 15, 424, 1961.
8. **Ramble, W. C. and Byrnes, W. R.,** Effects of certain common brush control techniques and materials on game food and cover on a power line right-of-way, No. 2, *Pa. Agric. Exp. Stn. Prog. Rep.,* 135, 1955.
9. **Ramble, W. C. and Byrnes, W. R.,** Ecological aspects of brush control — a long term study on a utility right-of-way, *Purdue Univ. Agric. Exp. Stn. Res. Bull.,* 835, 1, 1967.
10. **Ramble, W. C., Byrnes, W. R., and Hutnik, R. J.,** Effects of chemical brush control upon game food and cover, *Pa. Agric. Exp. Stn. Prog. Rep.,* 188, 1958.
11. **Ramble, W. C., Byrnes, W. R., and Worley, D. P.,** Effects of certain common brush control techniques and materials on game food and cover on a power line right-of-way, No. 3, *Pa. Agric. Exp. Stn. Prog. Rep.,* 151, 1956.
12. **Ramble, W. C., Byrnes, W. R., and Worley, D. P.,** Effects of certain common brush control techniques and materials on game food and cover on a power line right-of-way, No. 4*Pa. Agric. Exp. Stn. Prog. Rep.,* 175, 1957.
13. **House, W. B., Goodson, L. H., Godberry, H. M., and Duchter, K. W.,** Assessment of Ecological Effects of Extensive or Repeated Use of Herbicides, Midwest Research Institute, Kansas City, Mo., 1967.
14. **Pimentel, D.,** Ecological Effects of Pesticides on Non-target Species, Executive Office of the President, Office of Science and Technology, Washington, D.C., 1971.
15. Hazardous Materials Advisory Committee, Herbicide Report, Chemistry and Analysis of Environmental Effects on Agricultural and other Applied Uses, U.S. Environmental Protection Agency, Washington, D.C., 1975.
16. U.S. Department of the Interior, Management of Transmission Line Rights-of-Way for Fish and Wildlife, Biological Services Program FWS/OBS-79/22, U.S. Department of the Interior, Washington, D.C., 1979.

Chapter 16

AGRONOMIC SIGNIFICANCE OF RIGHT-OF-WAY CONTROL

INTRODUCTION

There is no one herbicide available that meets all requirements for complete control of vegetation, in the right-of-way or for the control of all vegetation. There are two major reasons. First, there are certain weed species that are resistant to each of the herbicides available; second, these herbicides do not behave equally well under all environmental conditions. To be effective, a soil sterilant must be soluble enough to be carried into the root zone by soil water; also, it must remain long enough for a lethal dose to be adsorbed by the plant. Aside from the relative solubility of the chemical, its movement and persistence in the soil are influenced by

1. The rainfall pattern of the area
2. The physical and chemical properties of the soil, such as texture, structure, and pH
3. The microorganisms in the soil that are able to deactivate the chemical

Failure to get expected control may result from an incomplete distribution of the herbicide in the soil or from amounts in the soil solution inadequate to effect a kill. A chemical of low solubility may be adsorbed in the upper soil layer and not reach deeper roots. This may happen in soils with poor underdrainage, in dry regions, or when the treatment was made at the wrong time. A soluble chemical may leach out of the soil before plant roots absorb a lethal dose. This occurs most commonly in sandy soils with excessive underdrainage and where rainfall is high, in seasons of heavy showers, and under irrigation. It may also result from improper timing of the treatment. When hard-to-kill species are a problem or where environmental factors reduce the activity of a chemical,

1. Increase the rate of application
2. Use of mixture of chemical
3. Repeat the treatment

ALL VEGETATION

The rates of application and the optimum time for treatment also vary with soil, rainfall, and the weed species to be controlled. In most areas, it is better to make repeated annual applications of soil sterilants at relatively light rates than to rely on a single heavy treatment. Such a maintenance dosage results in a smaller annual expenditure than a "one-shot" method and keeps chemicals in the surface soil where they can kill weed seeds coming in from outside. Nongrass species are usually the first vegetation to reinfest sterilized areas. Relatively inexpensive treatments with esters of 2,4-dichlorophenoxy (2,4-D) at 1 lb/acre will maintain areas free of vegetation for several additional years. Spot treat difficult-to-treat plants by hand rather than with a general spray treatment at rates high enough to kill the tolerant weeds.[1-4]

Where perennial weeds are susceptible to the phenoxy herbicides, it is economical to use a relatively insoluble soil sterilant like simazine or diuron to control annual weeds and 2,4-D or Silvex® to kill the perennials. The soil sterilant is applied preemergency to moist soil at 4 lb/acre in water. The phenoxy compound is applied to perennials at 1 lb/acre when they are in full leaf and growing rapidly to induce the maximum translocation to their roots. For this method to be successful, the soil sterilant must remain near the surface and there must be enough rain to activate it. Heavy rain will leach the herbicide and shorten the period of control. Incorporation of the chemical with the soil helps activate it in dry seasons, but this does not substitute for rain.

Industrial Sites

Soil sterilants prevent the growth of green plants. The soil may or may not be sterile. These chemicals are used in storage areas, lumberyards, and parking lots, on tennis courts and racetracks, under pipelines and transformer cages, under guardrails and around signposts on highways and lights on runways, near fire hydrants, trestles, and bridges, on public utility poles, and tank farms, along fence rows, for firebreaks, and in similar areas where any plant growth is undesirable. There are four major problems in maintaining bare ground:

1. No herbicide kills all species at reasonable rates of application
2. Reinfestation results from weed seeds in the soil after the herbicide has been leached below the surface
3. Desirable plants in adjoining areas may be killed or injured if surface water drains across treated areas to untreated ones or if tree roots extend into treated areas
4. Bare soil is subject to serious erosion where the terrain is steeply sloped and there is sufficient rainfall to cause runoff

Meet these problems by choosing the appropriate herbicide for the situation, and use satisfactory soil binding, stabilizing treatments, or structures to prevent erosion.

Paved Highways

Vegetation that encroaches upon the edges of concrete or asphalt pavement or grows up through cracks and holes causes premature breakdown of the pavement. Control with presurface and postsurface applications of herbicides is possible. Shoulders immediately adjacent to the trafficway, medians separating divided highways, and islands at highway intersections are often surfaced with asphalt. On such areas, apply herbicide to the gravel base just before it is ''shot'' with asphalt. A standard highway watering truck can be adapted for this use by equipping it with a loading pump (to circulate the spray material) and standard asphalt nozzles that deliver a fan spray.

Several herbicides prevent the emergence of plants through the pavement, but they vary in cost and in injury to vegetation adjacent to the paving. Plant growth on the unpaved area, particularly on fill slopes, may be desirable to prevent erosion.

Postpaving treatments will be necessary later to prevent encroachment from unpaved areas and to control vegetation growing up through the cracks in old pavements. Soil sterilants, including bromacil, diuron, monuron, and trichloroacetic acid (TCA), may be used to prevent reinfestation of treated areas.

Care should be exercised to prevent damage to desirable vegetation when using bromacil, monuron, and borate-chlorate. Bromacil, diuron, and monuron will injure trees adjacent to treated areas.

Railroads

Three distinct railway areas require weed control: the ballast, the roadbed, and the right-of-way. The ballast is a 12- to 16-ft-wide strip, made up of coarse material, such as cinders and gravel, that should be kept free of weeds. Because it is porous, it does not retain chemicals well. Insoluble herbicides — those absorbed through the leaves — and contact herbicides are most suitable. The roadbed (berm) beyond the ballast requires weed control, but elimination of vegetation increases erosion. The rest of the area to the right-of-way fence is similar to roadsides. If control is effected during the first 2 years by heavy rates of application, it can be maintained then with reduced rates. The effectiveness of some soil sterilants like bromacil, diuron, and simazine may not show up until the second or third year of use, especially in dry areas or with deep-rooted weeds.

CONTROL OF WOODY PLANTS

Noncropland control of woody plants — trees and shrubs — is quite important for general purpose activity. On militiary reservations, trees and shrubs must be removed for a clear line-of-sight on small-arms and artillery ranges and from around ammunition magazines. Woody plants must also be kept under control along highways, under powerlines, along fence rows, and at the ends of airport runways.[1-4]

Some of this growth can be cut mechanically; where this is not possible or practical, chemicals may be applied from the air or the ground using either sprays or pellets. Both selective and nonselective herbicides for killing woody plants are available and are applied in various ways, depending on the plants to be treated and their geographical location.

Foliage and Soil Applications

Both foliage and soil applications of herbicides may be used for the general control of woody plants.

Larger Trees and Brush

When trees and brush have become rather large, special control measures are necessary. These include basal-bark and stump applications and tree injections. In other cases, cut-surface applications, such as frills and girdles and cups and notches, are best.

Applications Along Roadsides and Utility Lines

Herbicides, if properly applied, are useful in the maintenance of roadsides and utility lines. The greatest dangers come from drift, runoff, improper application, and leaching to roots of desirable species under the treated area.

Poison Ivy, Poison Oak, and Poison Sumac

There are four satisfactory herbicides for killing poison ivy, poison oak, and poison sumac, i.e., amitrole, ammonium sulfamate (AMS), 2,4-D + 2,4,5-T, and 2,4,5-trichlorophenoxy (2,4,5-T). Apply when leaves are fully expanded and wet the foliage to the point of runoff. Drench the stems as high as possible when the plants are growing on a wall and allow the excess spray to run down to the roots. Plants growing in the shade require more amitrole or 2,4-D + 2,4,5-T mixture than those in the sun.

Retreatment, After an Initial Clearance

1. Oaks and maples — Use basal spray on clumps for maximum fill and minimum regrowth, follow with basal sprays on nonsuckering species or stem-foliage sprays on suckering species; or, use large volumes of stem-foliage spray; follow in 2 to 3 years with a basal spray.

2. Suckering species — Use overall stem-foliage spray. On susceptible species, repeat the following year; on resistant species, use basal sprays as followup.

For foliage applications, use esters of 2,4-5-T and 2,4-D on most species. Silvex® is better on some oaks and locust. Use amitrole for white ash, black locust, poison-oak, and poison-ivy. Use AMS in areas where drift from 2,4-D or 2,4,5-T is too hazardous. It can be used successfully in a mist blower in areas where ordinary application methods are difficult.

Make two complete sprayings of all transmission lines at 2-year intervals—the first to kill as much growth as possible and the second to kill escapees and resistant species. Spray only those species that grow tall enough to interfere with the lines. If they are over 6 feet tall, cut and spray the stump.

Fenuron pellets at 12½ lb/acre, active ingredient (ai), kill alder, blackberries, and sumac; elderberry, elm, hawthorn, maple, or willow are not killed, but all species are defoliated.

Picloram is an extremely potent and versatile herbicide available in liquid and pelleted formulations and in mixture or combination with several other herbicides. For certain hard-to-kill woody species, such as Macartney rose, persimmon, and salix, picloram is more effective than 2,4-D or 2,4,5-T, but has the same general herbicide activity.

Picloram has a very low toxicity to animal life and a very high level of toxicity to succulent and woody plants. Pasture and range grasses tolerate moderate level without injury. The chemical is relatively persistent in the soil. Special caution must be used during application to avoid drift and contamination of water supply used for irrigation or adjacent areas of susceptible crops and ornamental plantings.

PERSISTENCE OF HERBICIDES

Several factors are known to affect the fate of herbicides. These may include decomposition by microorganisms, leaching, temperature, and photochemical alterations. There are other processes that enter into the picture, such as adsorption, volatilization, chemical reactions, and adsorption and metabolism by the plant.

The wide range of herbicide persistence in soil is mainly related to their susceptibility to chemical decomposition. One of the factors contributing to their destruction is the action of soil microorganisms. This field of investigation has been pursued by a host of workers since the later 1940s. The previous section devoted to residues and persistence of the herbicides in soils reflects in many instances the results of the action of microorganisms which were responsible for the destruction of the herbicide.

Audus[5] has reviewed the evidence for microorganism breakdown, relationship of chemical structure to ease of attack by microorganisms, and the biochemical pathways and herbicide degradation. In addition, he as well as Newman and Thomas[8] reviewed the effects of herbicides on soil microorganisms. More recent reviewers have dealt with the effect on nontarget organisms[7] and environmental effects.[8] A summary of some contact sterilant and translocated herbicide for noncropland is given in Table 1.

Contact Herbicides

Sund[9] indicated that the ultimate breakdown of amitrole (3-amino-1,2,4-triazole) was due to microbiological attack. The rates of decomposition of amitrole in steam-sterilized soil are much lower than in unsterilized soils, indicating that decomposition is primarily due to the action of microorganisms.

Burschel and Freed[10] found that at 29°C all of the amitrole disappeared from the soil by the 7th week, and it had fallen from its original concentration to 50% within 4

Table 1
PERSISTENCE OF SOME NONCROPLAND HERBICIDES IN SOILS

Herbicide	Rate (kg/ha)	Time (months)	Ref.
Amitrole	3 — 10	1 — 3	1, 2
	10 — 12	4 — 5	3
Dalapon	7 — 15	1	2, 10
	15 — 20	3 — 4	11, 12
TCA, DSMA, cacodylic acid	15 — 30	4	17
Atrazine	2 — 4	4 — 8	4, 6
	3 — 8	8 — 12	7, 9
Simazine	2 — 5	12	12, 13
Diuron	1 — 2	4 — 8	3
	3 — 4	5 — 7	13
Monuron	2 — 4	5 — 6	13, 16
2,4-D	2 — 4	1 — 2	20
2,4,5-T	2 — 4	1 — 2	20
Picloram	0.1 — 0.2	18 — 24	21

weeks. The rate of recovery of 3-amino-1,2,4-triazole from Hagerstown silt loam soil was a function of temperature, pH, soil moisture, and microbial activity. There was a direct correlation between soil moisture and the persistence of amitrole. At the end of 6 days, 58% of the added amitrole was recovered from air-dried soil. In the same length of time, only 8% was recovered from soil containing 15% moisture, and none was recovered from soil containing 30% moisture.

Results from the laboratories of Burschel and Freed[10] showed the same trends. Amitrole dropped to 30% of the original concentration by 6 weeks at the 29°C temperature, whereas 48% of it was retained at 15°C. A broad range of factors (concentration, pH, microbiological toxicity, and organic matter) influenced the breakdown of amitrole. Increase in soil organic matter and temperatures up to 30°C increased the degree of microbial attack.

The fate and activity of TCA in the soil has been observed.[11] In sandy soil, relatively little breakdown of TCA was observed; the loss was proportional to the temperature. Much more breakdown occurred in silt loam and especially in muck; here again dissipation was the function of the temperature.

Kratochvil[12] studied the effect of TCA (10, 50, 100, 150 lb/acre) and endothall (1, 2, 4, and 8 lb/acre) on soil microorganism activity. All the rates were calculated on the active ingredient basis. The technique used involved the placing of soil treated with the chemicals in 250 Erlenmeyer flasks. Water content was adjusted to 35% by weight. At the end of 48 hr of incubation, the flasks were closed and incubation continued for 4 hr. At the end of that period, the pressure that had developed in the flasks was measured by a mercury manometer. Gas was generated by all levels of TCA; however, none of the pressures were as high as that generated on the check sample. A total of 10 lb/acre of TCA give a reading of about 7 cm of mercury, while the 150-lb/acre level of TCA produced only 2 cm of mercury pressure. The measurement of microbial decomposition with endothall was a considerably different picture: the levels of 1, 2, and 4 lb exceeded the check sample, and 8 lb only slightly depressed the pressure below that of the check sample. The herbicide TCA and dalapon disappear rapidly from the soil, they seldom persist more than a month or 6 weeks when they are used at the recommended levels of weed control.

Ehman and Birdsall[13] reported that cacodylic acid has an extremely low vapor pressure, and its translation in the soil is very slow. Disodium methanearsonate (DSMA) equivalent to 28 lb/acre was applied to the top of soil column and leached with 60 in. of water; less than 10% of the applied compound came through a 12-in. column of sand. Another experiment performed with 15 lb/acre of cacodylic acid in the same experimental setup and using an extrapolation of 60 in. of leaching water gave a leaching rate of 9% for sand and 6% for sandy loam.

Translocated Herbicides

The most commonly studied herbicide in this group is 2,4-D. Early investigators were concerned with persistence as related to moisture, temperature, and autoclaving. DeRose and Newman[14] were the first to report that the rates of application did not affect the persistence of either 2,4-D or 2,4,5-T; also, they reported that 2,4,5-T persisted longer than 2,4-D. Increased moisture levels in the soil led to high rates of decomposition, whereas persistence varied inversely with soil temperature (range 10 to 30°C).

The type of soil as well as differences in temperature as they affect 2,4-D decomposition was examined by Ogle and Warren.[15] The 2,4-D application rates were 2 and 4 lb/acre, respectively. The treated soils were placed in dishes in temperature chambers at 8, 12, 25, and 35°C. In general, the disappearance of 2,4-D was proportional to the temperature. Breakdown was more rapid in silt loam than in sandy soils and most rapid in muck. 2,4-D at 4 lb/acre disappeared rapidly in the muck soil, 1 week at 18 and 25°C and 4 weeks at 8 and 35°C.

Hartley[16] discussed the physicochemical aspects of the availability of herbicides in soil. He stated that the leaching of herbicides is affected by the solubility and the absorption of the herbicides in the soil, the moisture level of the soil at the time of application, and the amount of evaporation. Wiese and Davis[17] studied the movement of dimethylamine salts of 2,4,6-trichlorobenzoic acid (2,3,6-TBA), polychlorobenzoic acid (PBA), the amine salt of 2,4-D, the sodium salt of 2,3,6-trichlorophenylacetic acid (fenac), ester of fenac, 3-phenyl-1,1-dimethylurea (fenuron), 3-*p*-chlorophenyl) 1,1-dimethylurea (monuron), the amine salts of 2-(2,4,5-trichlorophenoxyl) propionic acid (Silvex®), and 2,4,5-T. The above herbicides were applied in water solutions or suspensions to tubes of soil. The depth of leaching in inches were as follows:

Dimethylamine salts of 2,3,6-TBA and TBA	18 and 19 ½
Amine salt of 2,4-D and sodium salt of fenac	15
Ester formulation of fenac, fenuron amine salt of silvex®, amine salt of 2,4,5-T, monuron	9
2,4,5-T, 2,4-D	3

In general there was a relationship between the depths of movement in the tubes and the solubility of the herbicides, except for perhaps the amine salt formulation of 2,4,5-T and Silvex®. Other studies have been performed utilizing the tube technique.[18] The 2,4-D amine salt was applied at the rate of 2 lb acid equivalent per acre on seven soil types. A given amount of rain was applied, and the study confirmed that the greatest movement occurred in the light sandy soils. A considerable movement occurred in granular types of clay soils. Further studies were undertaken using three formulations of herbicide on four sandy soils and one granular type. 2,4-D amine salt and 2,4-D isopropyl ester were compared, and there was virtually no difference in their pattern. 2,4-D ester, a 2,4-D amine salt, 2,4-D acid amide, 2,4-D acid and 2-methyl-4chlorophenoxyacetic acid (MCPA) potassium salt, and an MCPA acid were compared at three levels of rainfall, ½, 1, and 2 in. Surprisingly, the acids and the 2,4-D acid amide, although relatively insoluble, move appreciably.

A comparative study of the effects of 2,4-D and picloram has been made on a specific strain of *Aspergillus niger* V.T.[19] Levels of 0, 0.4, 2, 10, and 50 ppmw were used in the culture studies. 2,4-D significantly reduced growth of *A. niger* at concentrations of 10 and 50 ppmw in cultural media. Growth depression of *A. niger* was not apparent with picloram. Highly significant decreases in bean plant growth were obtained with all herbicide levels. Goring and associates[20] tested the effect of picloram on 45 common organisms. The incubation studies were for 1 day, 1 week and 1 month. Picloram at concentrations of 1000 ppm did not seem to have significant gross effect on the fungi and bacterial population of the soil. Doses as high as 1000 ppm in soil did not appear to significantly affect CO_2 evolution, urea hydrolysis, or gross counts of bacteria and fungi.

The decomposition of picloram has been studied[21] over a temperature range of 34 to 95°F, utilizing two soil types designated as A and B. The composition of soil types A and B were as follows:

	Sand (%)	Silt (%)	Clay (%)	Organic matter (%)	pH
Type A	59	35	6	1.5	5.7
Type B	29	43	28	3.2	6.2

As indicated, 10- to 25-fold increases in rate of decomposition were obtained by increasing the temperature from 35 to 94°F.

Hamaker and associates[22] compared the leaching rates of 2,4-D, 2,4,5-T, and picloram. These three chemicals were applied to the top of columns containing sandy silt loam at the rate of about 100 lb acid equivalent per acre. These columns were leached with 2, 4, and 8 in. of water. They used radish plants as an indicator and found that in all cases, columns leached with 2 in. of water, produced symptoms in the radishes to a depth of 12 in. or more. When they were leached with 9 in. of water, the chemicals were moved so far down the column that no symptoms were observed in the radish plant until a depth of at least 10 to 12 in. It would appear from these tests that picloram is like 2,4-D and 2,4,5-T in that it is easily leached under high rainfall conditions.

Herr and associates[23] obtained a broader view of the leaching capacity in picoloram using a wider range of soil characteristics, silty clay, a silt loam light-textured, and a silt loam medium-textured.

Surface applications of picloram were made on plots at the rates of 0, 2, 4, 8, 32 and 64 oz/acres at each location. The amount of precipitation that occurred from the time picloram application was made, to the sampling of the silt loam soil, was 2.5, 15.8, and 26.7 in., respectively. Corresponding data on the light-textured silt loam were 2.3, 24.9, and 41.8 in. The amount of precipitation that occurred in the medium-textured silt loam during these periods was 1.9, 27.3, and 45.9 in. In the silty clay soil at the 280-day level, the amounts of picloram in the first 6 in. of soil was about one half of that obtained on the 101st day for the 2,4, and 8 oz/acre treatment. For the 64-oz/acre treatment, at the end of 476 days, there was still a high concentration in the first 18 in. and most of it was in the top 6 in. The results on the light-textured silt loam indicated that for the 2-, 4-, 8-, 32-, and 64-oz levels by the end of 245 days, that a considerable amount of leaching had taken place down to the 24-in. level. On the medium-textured silt loam; much more of the picloram had leached out of the upper levels at 2-, 4-, and 8-oz applications per acre than on the silty clay. However, at the high levels, 64 oz/acre, at least one third of the original amount in the first 6 in. was still present and amounts up to 78 ppb had leached down to the 30-in. depth. The data indicated that picloram residues moved faster and more completely in the lighter texture of soil with a low organic content. This trend is also reported by Goring and associates.[21] Substantial amounts of picloram were found more than 430 days after application of 32 and 64 oz/acre on all three soils, indicating residues from larger applications were quite persistent.

In these photodecomposition studies,[24] 1 mg of picloram was spread evenly over the surface of an 8-cm petri dish or on soil placed within the petri dish. The source of UV light was a mineral light (model R-51 lamp) which provided approximately 155 W of shortwave UV radiation per square centimeter. Picloram content was determined after 48- and 168-hr exposure to the light conditions. It was found that the picloram was deactivated more by UV light than by sunlight. By chromatographic analysis, 60% of 1 mg of picloram applied to petri dishes was degraded by UV light after 48 hr, whereas 35% was degraded by sunlight. After 7 days' exposure, more than 90% was degraded by UV light and 65% was degraded by sunlight. In soil, only 15% of the picloram was degraded by exposure for 7 days to sunlight.

Soil Sterilants

Atrazine and simazine are adsorbed to certain soil constituents. Even with continuous application for years, atrazine is rarely found below 12 in. in the soil. The greater portion of simazine will be found in the upper 2 in. of soil. Simazine has a low water solubility. The decomposition of s-triazine herbicides by soil microorganism has been studied by several investigators.

Kaufman et al.[25] developed a population of soil microorganisms that effectively degraded simazine by using enrichment techniques. They found that degradation would begin immediately, but it continued very slowly in that only 3 to 4% was degraded in 30 days. One organism, *Aspergillus fermigatus* Fres., could degrade simazine almost completely in 12 days. Simazine was degraded in soil in about 135 days to less than 10% of its original concentration at a temperature of 25°C. At 18°C a soil containing 1% humus was degraded about 50% in 150 days, and in a soil containing 10% humus, decomposition had taken place to about 85% in 170 days.

Chandra et al.[26] examined the effects of high concentrations of simazine on microorganism activity in nine Oregon soils. This study was conducted concurrently on S-ethyl dipropylthiocarbamate (EPTC), diuron, and 2,3,6-TBA. A total of 72 treatments were evaluated by carbon dioxide evolution methods. The 72 treatments included 9 soils, 4 herbicides, and 2 treatment levels, 5 and 100 ppm. The 100-ppm level is 20 to 35 times normal rate—field practice. Although there were obvious differences between soils in their response, the herbicides in most cases depressed carbon dioxide evolution for at least 28 days. For only 2 treatments was the depression significantly greater at 56 days. The authors felt the decrease in effect after 28 days could indicate

1. Acquirement of microbial tolerance for the herbicides
2. Their decomposition
3. Their possible inactivation by adsorption on organic matter of clay

In most instances, the depression of carbon dioxide evolution was greater with the 100-ppm level than at the 5-ppm level. Diuron showed different interactions with the different soil types at either rate, it appeared to depress carbon dioxide production most at 14 days. In a few instances, notably with simazine in 2 soil types at 7 and 14 days, carbon dioxide evolution was temporarily increased 12 to 14% by herbicide treatment at the 5-ppm level. Higher rates of the herbicides did not inhibit in proportion to the concentrations applied to the different soils. Generally, chemicals that are persistent are normally those that resist leaching and are retained in the surface layers of the soil.

The effect of UV light has been studied on a number of herbicides (substituted ureas, triazines, bipyridyl compounds, uracils, etc.). Atrazine and simazine and ametryne will

Table 2
SPECIFIC WEED PROBLEMS AND TREATMENTS FOR NONCROPLAND CONTROL

Broadleaf herbaceous weeds (selective control in desirable grasses)

Problem: Annuals, seedling, or shallow-rooted perennials

Treatment:

No. 1. Dicamba at 0.5 to 1 lb/acre on early weed growth in spring (Remarks: Kills only top growth of many established perennials)

No. 2. 2,4-D at 1 to 2 lb/acre on early growth to first bloom (Repeat as necessary to maintain control during growing season.)

No. 3. 2,4,5-T at 1 to 2 lb/acre on early growth to full bloom on weeds difficult to control with 2,4-D, such as nightshade

No. 4. Silvex at 1 to 2 lb/acre on early growth to full bloom on weeds difficult to kill with 2,4-D, such as black medic, chickweed, clover

Problem: Deep-rooted and other hard-to-kill biennials and perennials

Treatment:

No. 1. Dicamba at 0.25 to 2; 4 to 8 lb/acre when weeds are in active growth stage (The highest rates may injure grass temporarily and are effective as soil treatments made before periods of precipitation.)

No. 2. Picloram at 0.5 lb/acre on very active vegetative growth to early bloom in spring or summer during vigorous rosette growth, or as soil application in the fall (Precaution: Use special care to avoid spray drift on desirable plants or contamination of irrigation water; apply only as thickened spray; picloram persists for several years in some soils.)

No. 3. Picloram plus 2,4-D mixtures (follow instructions on label) (Precautions: same as for picloram used alone)

No. 4. 2,3,6-TBA at 10 to 20 lb/acre sprayed on succulent foliage, preferably during period of adequate rainfall (Precaution: some desirable grasses injured by high rates)

No. 5. 2,4-D at 2 to 4 lb/acre sprayed on vegetative growth to early bloom in spring or summer or on vigorous rosette growth in fall. (Repeat as necessary to eliminate or reduce stand.)

No. 6. 2,4,5-T at 2 to 4 lb/acre. Use same as 2,4-D, but on weeds resistant to 2,4-D, such as perennial horsenettle (Precaution: Do not use where drift or runoff may contaminate water or apply sprays in close proximity to residential sites.)

Weedy grasses and sedges where control of broadleaf weeds is not necessary

Problem: annuals and seedling perennials

Treatment:

No. 1. Aromatic oil at 40 to 80 gal/acre sprayed on young growth (Repeat as necessary to maintin control.)

No. 2. Dalapon at 5 to 10 lb/acre sprayed on new growth 4 to 8 in. tall (Repeat as necessary to maintain control.)

No. 3. Dinoseb-fortified fuel oil at 40 to 80 gal/acre sprayed on new growth (Use 2 to 3 pt dinoseb per 100 gal fuel oil.) (Precaution: do not contaminate water to be used for livestock watering or domestic purposes.)

Problem: established perennials (bermudagrass, johnsongrass, phragmites, paragrass, quackgrass, reed canarygrass, and sedges)

Treatment:

No. 1. Amitrole at 4 to 10 lb/acre sprayed on new growth before heading (Repeat two to three times each season to eliminate or reduce stand. Lower rates will suppress growth.) (Precaution: do not contaminate water to be used for livestock watering or domestic purposes.)

No. 2. Amitrole-T at 4 to 8 lb/acre used the same as amitrole

No. 3. Aromatic oil at 120 to 160 gal/acre sprayed on new growth before heading (Repeat every 3 to 4 weeks during growing season to eliminate or reduce stand.)

No. 4. Dalapon at 10 to 15 lb/acre sprayed on new growth (Repeat as necessary to suppress growth.)

No. 5. Dalapon at 20 to 30 lb/acre sprayed on early growth to early heading (Repeat every 6 to 8 weeks or as necessary to eliminate or reduce stand.)

No. 6. Dinoseb-fortified fuel oil at 120 to 160 gal/acre sprayed on new growth (Use 2 to 3 pt dinoseb per 100 gal fuel oil. Repeat every 3 to 4 weeks during growing season.)

Table 2 (continued)
SPECIFIC WEED PROBLEMS AND TREATMENTS FOR NONCROPLAND CONTROL

No. 7. DSMA at 2.5 to 7 lb/100 gal of water (Treat when weeds are actively growing.) (Remarks: Spray plants until thoroughly wet. Add 2 to 3 pt of surfactant per 100 gal water.)

No. 8. MSMA or cacodylic acid at 2 to 5 lb/acre on actively growing weeds (Remarks: Add 1 to 2 qt of surfactant and spray plants until thoroughly wet.)

No. 9. TCA (sodium) at 20 to 40 lb/acre sprayed on dead top growth of reed canarygrass after freezes in late fall (Follow with amitrole-T at 4 lb/acre in midsummer to kill canarygrass seedlings. (Remarks: Encourages establishment of redtop, bluegrass, and other desirable low-growing grasses.)

No. 10. TCA (sodium) at 50 to 200 lb/acre sprayed on aftermath growth in fall or as soil treatment (Remarks: More effective when adequate precipitation follows; at lower rates, desirable grasses may be seeded next season.)

Mixed herbaceous and grass weeds

Problem: Annuals, biennials, shallow-rooted perennials, and seedling perennials (lower rates); established deep-rooted and hard-to-kill perennials (higher rates)

Treatment:

No. 1. Amitrole-T at 3 to 5; 6 to 12 lb/acre (Spray on foliage at early growth to first bloom. Repeat as necessary to maintain control or eliminate stand.) (Precaution: Do not contaminate water to be used for irrigation, domestic use, or livestock watering.)

No. 2. Amitrole plus simazine mixture (follow directions on label) (Remarks: Give longer-lasting control than amitrole alone.) (Precaution: Do not contaminate water to be used for irrigation, domestic purposes, or livestock watering.)

No. 3. AMS at 50 to 100 lb/100 gal of water (Spray on active growth. Mix 0.5 to 2 lb/gal water and spray to wet foliage.) (Remarks: Effective on woody and herbaceous plants. One formulation registered for use adjacent to aquatic areas including domestic water supplies.)

No. 4. Aromatic oil at 120 to 160 gal/acre (Spray on new growth. Repeat as necessary or every 3 to 4 weeks to eliminate or reduce stand.)

No. 5. Borate (sodium) at 30 to 120 lb/1000 ft^2; 1300 to 5200 lb/acre (Apply to active or dormant growth. Precipitation required after application.)

No. 6. Borate mixtures with sodium chlorate or other chemicals or both (follow instructions on label) (Spray on active growth or make soil applications before period of adequate precipitation.) (Remarks: Commonly used where soil sterilization (bare soil) is desired.)

No. 7. Bromacil at 3 to 6; 10 to 25 lb/acre (Before or during early growth of weeds.) (Remarks: Apply higher rates before period of greatest precipitation. Adequate precipitation required following application.)

No. 8. Chlorate (sodium) at 300 to 600; 700 to 1300 lb/acre. (Spray foliage or make soil applications before period of adequate precipitation.) (Precaution: Observe all precautions to avoid fire hazard.)

No. 9. Chlorate mixtures with borates or other chemicals (follow directions on label) (Use same as chlorate alone.) (Remarks: Fire hazard usually less than chlorate alone but caution is necessary.)

No. 10. Dalapon at 5 to 10 or 15 to 30 lb/acre plus 2,4-D or silvex at 1 to 2 or 2 to 4 lb/acre (Spray early growth to early bloom. Use silvex where broadleaf weeds or brush resistant to 2,4-D are present.)

No. 11. Diuron or monuron at 5 to 20 lb/acre; 20 to 60 lb/acre before period of adequate precipitation (Remarks: Diuron gives longer soil sterility; monuron is more effective on deep-rooted weeds.)

No. 12. Dinoseb-fortified fuel oil at 40 to 80; 120 to 160 gal/acre on young growth to first heading (Repeat as necessary to maintain control or every 3 to 4 weeks to eliminate or reduce stand; use 2 to 3 pt dinoseb per 100 gal fuel oil.)

No. 13. Erbon at 120 to 160 lb/acre when weeds are young and succulent or during period of adequate precipitation (Remarks: Make soil applications in late fall or winter in regions of wet winters and dry summers.)

No. 14. Ethylene glycol *bis* (trichloroacetate) at 10 to 20 lb/acre (Spray when weeds are about 12 in. tall.) (Remarks: Apply in fuel oil in sufficient volume to completely cover foliage.)

No. 15. Monuron-TCA at 20 to 40 lb/acre; 40 to 66 lb/acre when weeds are young and succulent or during period of adequate precipitation (Remarks: Make soil applications in late fall or winter in regions of wet winters and dry summers.

Table 2 (continued)
SPECIFIC WEED PROBLEMS AND TREATMENTS FOR NONCROPLAND CONTROL

No. 16. Oil at 40 to 50 gal/acre aromatic or fortified fuel, plus 2,4-D ester at 2 to 3 lb/acre on young growth up to first heading (Apply same as for aromatic or fortified fuel oil.) (Remarks: More economical and often more effective than oil alone.)

No. 17. Paraquat (cation) at 0.5 to 2 lb/acre (Spray young growth. Repeat as necessary to maintain control.) (Remarks: Kills top growth only of perennials.)

No. 18. Prometone at 10 to 15 lb/acre; 20 to 60 lb/acre (Before weed emergence or during early growth. Mix in water or oil.) (Remarks: Oil mixture kills top growth quicker. Precipitation required for effective preemergence treatment.)

No. 19. Simazine or atrazine at 5 to 20 lb/acre; 20 to 40 lb/acre (Spray before or during early growth to first heading of weeds.) (Remarks: Simazine requires more precipitation for effective root kill; remains effective in soil longer.)

Undesirable woody plants

Problem: alder, baccharis, boxelder, cottonwood, poplar, sycamore, willow, and other species susceptible to 2,4-D

Treatment:

No. 1. AMS at 7 to 10 lb in 2 gal of water. (Remarks: Spray over freshly cut stump or in frills or notches cut around base of tree. Cut notches every 6 in. around circumference to tree; crystals may be sprinkled liberally over cut stumps or in frills or notches.)

No. 2. AMS at 60 lb/100 gal of water at full leaf in spring or summer. (Remarks: Thoroughly wet foliage; with airblast mist-spray equipment use 100 to 400 lb/100 gal of water and apply at lower volume; to apply in oil-water emulsion follow directions on label.) (Precaution: Use mist spray only in areas far removed from susceptible crops.)

No. 3. Fenuron (where precipitation exceeds 15 in. annually) at 12 to 25 lb/acre in spring from bud growth to full leaf and just before or during periods of adequate precipitation (Remarks: Apply broadcast or in grid pattern with 1 to 2 teaspoonfuls 3 ft apart; apply same rate around base of individual trees or clumps of brush for spot treatments. Increase rate 50% for cherry, dogwood, elderberry, osageorange, persimmon, and sassafras.)

No. 4. Fenuron-TCA at 12 to 25 lb/acre (Spray at full leaf stage or broadcast granular or pelleted formulations before or during periods of precipitation in spring.) (Remarks: Available in liquid, granular, and pelleted formulations.)

No. 5. Picloram at 6 to 8.5 lb/acre (Broadcast granules just before or during periods of precipitation.) (Remarks: Highest rate is necessary for ash, blackgum, and oak; for spot treatments under individual trees or clumps of brush broadcast 1 to 2 tablespoonfuls/30 ft^2 of soil surface. Picloram persists for 3 years in some soils.) (Precaution: Do not use near ornamentals or other desirable plants and avoid treating areas where runoff may reach cropland or irrigation water.)

No. 6. Picloram plus 2,4-D mixture (follow instructions on label). Spray at full leaf during vigorous growth (Remarks: Spray solution may also be applied to cut stumps or in frills; apply only in thickened spray; picloram persists in some soils for 3 years.) (Precaution: Use extreme care to avoid drift of spray to nearby desirable plants, and do not contaminate water to be used for irrigation.)

No. 7. Silvex at 2 to 4 lb/acre (Spray at full leaf during active growth; spring treatments are best for some species, such as saltcedar.) (Remarks: Silvex is especially effective on maple, mulberry, palmetto, redbud, saltcedar, and trumpetvine.)

No. 8. 2,3,6-TBA at 10 to 20 lb/acre (Spray foliage and soil when conditions are suitable for active growth.) (Remarks: Especially effective on woody vines.)

No. 9. 2,4,5-T at 2 to 4 lb/acre (Spray at full leaf during active growth, spring treatments best for some species, such as salt-cedar.) (Remarks: Especially effective on brambles, mesquite, oak, osageorange.) (Precaution: Do not use where contamination of any type of water may occur. Do not use sprays in close proximity of susceptible crops or residential areas.)

Supplemental information

1. Rates are in terms of acid equivalent or active ingredient of herbicide which may or may not correspond to the same units of concentration of the formulation.
2. Some herbicides registered for use on noncropland may not be included in this guide.

Table 2 (continued)
SPECIFIC WEED PROBLEMS AND TREATMENTS FOR NONCROPLAND CONTROL

3. During the use of herbicides recommended for control of perennial grasses or sedges, the weakened stands are frequently invaded by broadleaved weeds. A total of 1 to 2 lb/acre of amine or low-volatile ester derivatives of 2,4-D added to the treatment for grass control will usually eliminate the broad-leaved weeds.

undergo photodecomposition, and there are differences in the changes in the absorption spectra for these herbicides when exposed to sunlight and UV light.[27-29]

A number of herbicides that undergo photodecomposition have been studied from the standpoint of the effect of far, middle, and near UV. The experiment included four substitutes, monuron, diuron, neburon, and fenuron; two triazines, simazine and atrazine; and two uracil herbicides, bromacil and isocil. For monuron, neburon, and fenuron, the most rapid loss occurred during the first 24 hr. A steady-state situation was reached with no further herbicide loss for these three chemicals. In comparing simazine and atrazine, the latter degraded most rapidly during the first 80 hr and continued to disappear at a slower rate. For bromacil and isocil, the loss was almost the same for the two compounds when irradiated with middle and far UV. More bromacil than isocil was lost under near UV.

SPECIFIC WEED PROBLEMS

Examples of various herbicide treatments that have been used successfully in noncropland are given in Table 2. These examples show the scope of herbicide usage in noncropland. They are also guides for the use of herbicides under local conditions. However, rate ranges rather than specific rates are given. Times of application are given in general terms because local climate, soil composition, slope of land, proximity to canals or ponds or desirably plants, and methods required for their safe use. Spot treatments for control of perennial weeds should be used where possible, especially where there is a possibility of contaminating water or injuring nearby desirable plants. Specific details for the safe and effective local use of herbicides can be obtained from weed specialists in the state agricultural experiment stations.

REFERENCES

1. Department of the Army, Herbicide Manual for Noncropland Weeds, Technical Manual TM 5-629, U.S. Government Printing Office, Washington, D.C., 1970.
2. Department of the Navy, Basic Pest Control Technology, Naval Facilities Engineering Command, Alexandria, Va., 1976.
3. Department of the Navy, Chemical Control of Disease Vectors and Economic Pests, Naval Air Station, Jacksonville, Fla., 1976.
4. **Stewart, L. E., Gienger, G. W., Parochetti, J. V., and Coan, R. M.,** Broadcast Spraying, Extension Bulletin 229, University of Maryland, College Park, Md., 1976.
5. **Audus, L. J.,** Herbicide behavior in the soil, in *The Physiology and Biochemistry of Herbicides,* Audus, L. J., Ed., Academic Press, London, 1964, 163.
6. **Newman, A. S. and Thomas, J. R.,** Decomposition of 2,4-dichlorophenoxyacetic acid in soil and liquid media, *Soil Sci. Soc. Am. Proc.,* 14, 160, 1949.
7. **Pimentel, D.,** Ecological Effects of Pesticides on Non-Target Species, Office of Science and Technology, U.S. Government Printing Office, Washington, D.C.

8. Hazardous Materials Advisory Committee, Herbicide Report, Chemistry and Analysis of Environmental Effects on Agricultural and other applied uses, U.S. Environmental Protection Agency, Washington, D.C.
9. **Sund, K. A.**, Residual activity of 3-amino, 1,2,4-triazole in soils, *Agric. Food Chem.*, 4, 57, 1956.
10. **Burschel, P. and Freed, V. H.**, The decomposition of herbicide in soils, *Weeds*, 1, 157, 1959.
11. **Ogle, R. E. and G. F. Warren**, Fate and activity of herbicides in soils, *Weeds*, 3, 257, 1954.
12. **Kratochvil, D. E.**, Determinations of the effect of certain herbicides in soil microorganisms, *Weeds*, 1, 26, 1951.
13. **Ehman, P. J. and Birdsall, J. J.**, Fate of cacodylic acid in soils and plants, Contract DA-18-064-CML-2836(A), Ansul Chemical Company for U.S. Army Biological Laboratories, Fort Detrick, Frederick, Md., 1963.
14. **De Rose, H. R. and Newman, A. S.**, The comparison of the persistence of certain plant growth regulators when applied to soil, *Soil Sci. Soc. Am. Proc.*, 12, 222, 1947.
15. **Ogle, R. E. and Warren, G. F.**, The fate and activity of herbicides in the soil, *Weeds*, 1, 257, 1951.
16. **Hartley, G. S.**, Physico-chemical aspects of the availability of herbicides in soils, in *Herbicides and the Soil*, Woodford, E. K. and Sager, G. R., Eds., Blackwell Scientific, Oxford, 1960, 63.
17. **Wiese, A. F. and Davis, R. G.**, Herbicide movement in soil with various amounts of water, *Weeds*, 12, 101, 1964.
18. **Hattingh, E. R.**, Survey of certain weed control problems in Africa south of the Sahara, in Proc. 4th British Weed Control Conf., 1958, 215.
19. **Arnold, W. R., Santleman, P. W., and Lynd, J. G.**, Picloram and 2,4-D effects with *Aspergillus niger* proliferation, *Weeds*, 14, 89, 1966.
20. **Goring, C. A. I., Griffith, J. D., O'Melia, F. C., Scott, H. H., and Youngson, C. R.**, The effect of Tordon on microorganisms and soil biological processes, *Down Earth*, 22 (4), 14, 1967.
21. **Goring, C. A. I., Youngson, C. R., and Hamaker, J. W.**, Tordon herbicide — disappearance from soils, *Down Earth*, 20 (4), 3, 1965.
22. **Hamaker, J. W., Goring, C. A. I., and Youngson, C. R.**, Sorption and leaching of 4-amino-3,5,6-trichloropicolinic acid in soils, *Adv. Chem. Ser.*, 60, 1966.
23. **Herr, D. E., Stroube, E. W., and Ray, D. A.**, The movement and persistence of picloram in soil, *Weeds*, 14, 248, 1966a.
24. **Tschisly, F. H.**, Response of Tropical and Subtropical Woody Plants to Chemical Treatments, (ARPA order 424), U.S. Department of Agriculture Technical Report, U.S. Department of Agriculture, Washington, D.C., 1967.
25. **Kaufman, D. D., Kearney, P. C., and Sheets, T. J.**, Microbial degradation of simazine, *J. Agric. Food Chem.*, 13 (3), 238, 1965.
26. **Chandra, T., Furtick, W. R., and Bollen, W. B.**, The effect of four herbicides on microorganisms in nine Oregon soils, *Weeds*, 8, 589, 1960.
27. **Jordon, L. S., Day, B. E., and Clerx, W. A.**, Photodecomposition of triazines, *Weeds*, 12, 5, 1964a.
28. **Jordon, L. S., Mann, J. D., and Day, B. E.**, Effects of ultraviolet light on herbicides, *Weeds*, 13, 43, 1965.
29. **Jordon, L. S., Coggins, C. W., Jr., Day, B. E., and Clerx, W. A.**, Photodecomposition of substitute phenyl ureas, *Weeds*, 12, 1, 1964b.

Part IV: Aquatic Rights-of-Way Management

Chapter 17

MANAGEMENT OF FLOATING, SUBMERSED, AND EMERSED AQUATIC VEGETATION*

INTRODUCTION

An aquatic weed can be simply defined as an aquatic plant that grows when and where it is not wanted. For example, spatterdock (*Nuphar advana*) offers a desirable habitat for various species of fish and is pleasing to look at, but a patch of spatterdock that clogs the area around a boat dock is a nuisance to those using the dock. Obviously, waters with a variety of users are likely to contain aquatic plants that are considered desirable to some groups and objectionable to others. Compromise, therefore, is necessary.

Aquatic weed control requires skill. All possible consequences of a program should be carefully weighed. The first step in control is to identify the problem plants. Along with identification, something should be known about the life history and growth pattern of the problem plants. Once the plant is identified and the problem is understood, control can be planned.[1-8]

METHODOLOGY

The basic methods for controlling aquatic plants are mechanical, environmental, biological, and chemical. They can be, and often are, used in combination.

Mechanical Control

Physical removal is most effective for small quantities of plants near shorelines (Figures 1 and 2). The technique consists of cutting, mowing, raking, digging, or pulling. In small ponds, lakes, or streams, plants like sedges, reeds, cattails, and rushes can be mowed with a sickle or scythe or dug out with a hoe. Small patches of water lilies or spatterdock may be cut or pulled and destroyed before extensive growth occurs. However, pulling or cutting must be repeated several times to eliminate all new growth as soon as it appears. Single treatments are usually not effective because of the extensive tubers and root systems of most aquatic plants.

Attempts to remove submersed plants by cutting or dragging a chain or cable over the bottom are not effective. Actually, this practice may cause spreading of the plants because each stem fragment left in the water is potential new plant. For larger areas, special cutting machines have been developed that can be adjusted to cut at various depths. Several cuttings are required each year.

In small ponds, thin black plastic has been used to eliminate plant growth. When plastic is placed on the bottom in early spring, it prevents the resumption of growth of rooted aquatics. This technique can be especially effective in private waters where special local control around boat docks or at the edge of weed beds is desired.

For large areas, such as reservoirs, flood control canals, irrigation ditches, etc., large-scale equipment is needed, and aerial spraying is often done.

Environmental Control

Deepening very shallow ponds or lakes to 10 ft or more offers the most permanent method of weed control. With man-made lakes and ponds, the cost of doing this may

* Abridged and updated from the Research Planning Conference on Integrated Systems of Aquatic Plant Control, Proceedings, October 29 to 30, 1973, U.S. Army Waterways Experiment Station, Vicksburg, Miss., 1974. With permission.

FIGURE 1. Recreation beach, Lake Francis Case, Pickstown, S.D. The marginal weedy area above the beach can be best controlled by grading and mowing at regular intervals. (Photograph courtesy of the U.S. Army Corps of Engineers, Omaha district.)

be high at the time the pond is constructed, but it is more expensive to dredge to this depth after the pond is built. At depths of 10 ft or more, it is unusual for submersed vegetation to reach the surface. Sunlight penetration at this depth usually is not sufficient to stimulate vigorous growth of most aquatic plants.

With reeds, sedges, rushes, and floating species such as water lilies, steep sides that drop off rapidly into deep water (5 ft or more) will usually reduce growth or limit it to the very edge of the pond, where the weeds may be cut easily with a scythe or by other mechanical means. However, steep sides may create other problems. If small children frequent a pond, steep sides may be a safety hazard.

Spawning of some species of fish requires shallow water and aquatic vegetation for successful reproduction. While protection from predation is provided small game fish, if aquatic plants are present, excessive plant growth can be harmful in the overprotection of undesirable species.

Fertilization of ponds or lakes in the early spring may stimulate the growth of plankton and change the environment. Fertilized waters become greenish or brownish in the warm season, reflecting the presence of millions of minute plants and animals. This "bloom" restricts the penetration of sunlight to the bottom, inhibiting photosynthesis and depressing the growth of submersed plants.

Biological Control

Animals that are dependent on aquatic plants for food are of limited value for the control of these plants (Figures 3 and 4). Muskrats may cut considerable quantities of plants, but in doing so, leave fragments that may act as sources for plant propagation. Waterfowl may utilize certain floating and submersed plants; however, the fertility resulting from their excrement may create excessive algal blooms that may be even less desirable than the original problem. Some exotic species of fish such as the white amur (*Ctenophoryngodon idella*) are being promoted to control aquatic vegetation.

Chemical Control

Herbicides offer a solution to many otherwise difficult aquatic weed problems (Figures 5 and 6). However, chemical control has its own special set of limitations. Most

FIGURE 2. Recreation area at Melvern Lake, Kansas. Playground areas of indigenous materials. Access roads and parking areas are bituminous. Camp pads are compacted with limestone. Weeds are controlled by mowing where possible and hand spraying with herbicides around the trailor pads and parking lots. (Photograph courtesy of the U.S. Army Corps of Engineers, Kansas district.)

unsuccessful chemical weed-control programs are not the fault of the herbicide. Failures are usually the result of errors in calculation or of the improper selection of chemicals for the weed species present. Inadvertent damage such as fish kills can usually be traced to errors in judgment, arithmetic, application of chemicals, or to oxygen depletion.

APPLICATION

Apply chemicals in accordance with the manufacturer's label instructions. These instructions generally include recommended equipment and dosage and suggested time of application; they are specific for the plant species to be controlled. Proper equipment will save time and labor and will permit uniform application. The chemical should be applied when the plant is most susceptible, and the dosage used should never exceed the amounts listed on the label. Only the active ingredient of most commercial herbicides is the effective controlling agent. In most instances the recommended dosage is based on the active ingredient. Therefore, one must know the percentage of active ingredient in the material before he can plan the application rate and before he can compare the cost of one herbicide with another.

Dosage Calculations

The need for strict adherence to proper procedure and accurate calculations cannot be overemphasized. Accurate dosage is essential, especially in view of the relatively small concentration of chemicals used in aquatic situations. The concentrations used are so low that they are most conveniently expressed as parts per million (ppm). For example, if a 2-ppm concentration of the active ingredient in a herbicide will kill an aquatic weed, it means that the concentration for the proposed treatment should be in a ratio of 2 parts of the active ingredient to 1 million parts of water.

In order to calculate the amount of chemical to be used, it is necessary first to know how much water is in your pond. If it was constructed under the direction of the Soil Conservation Service, you probably can obtain a very precise value for the volume of the pond. If these data are not readily available, it will be necessary to calculate the

A

FIGURE 3. Biological control of alligatorweed in Lake Langhofer, Pine Bluff, Ark. (A) Before release of *Vogtia malloi*, host specific for control of alligatorweed; (B) 1 year after release of the moth. Note nearly complete control of alligatorweed. (Photograph courtesy of Dr. P. C. Quimby (entomologist), U.S. Department of Agriculture, Stoneville, Miss.)

total volume from your measurements of surface area and average depth. Total volume in acre-feet can be obtained by multiplying the surface area in acres by the average depth in feet. For example, a 2-acre pond with an average depth of 4 ft contains 8 acre-ft of water. An acre-foot of water weighs 2.7 million lb. Therefore, a concentra-

FIGURE 3B

tion of 1 ppm can be obtained by dissolving 2.7 lb of the active ingredient (ai) in 1 acre-ft of water. The example pond has 8 acre-ft of water, so 21.6 (2.7 × 8) lb of ai would be needed for 1 ppm concentration. To get a concentration a ½ ppm in the 8 acre-ft pond, multiply 2.7 × 5 ×8. The product, 10.8, is the number of pounds required. The volume of small ponds may best be handled as cubic feet, computed by multiplying surface area in square feet by the average depth in feet. Thus, if a pond has a surface area of 5,000 ft^2 and an average depth of 4 ft, it would contain 20,000 ft^3 of water. Since 1 ft^3 of water weighs 62.5 lb, there would be 1,250,000 lb of water in the pond. To achieve a 1 ppm concentration, 1.25 lb of ai would be needed.

Hazards of Chemicals

Some herbicides are toxic to fish and other animals, but the amounts needed to kill aquatic weeds are generally much less than the tolerance level of fish and other animals. Newly hatched fish have less tolerance to herbicides than do older fish. In some instances, minor fish kills may occur when the safe dosage is exceeded due to an uneven distribution of the chemical.

Fish mortality may also occur as an indirect result of chemical weed control because of the oxygen consumed by the rapid decay of the destroyed plants. To minimize this danger, treat only one third of the water area at any one time, spacing applications at least 1 week apart so that rapid plant decay will not reduce the oxygen in the water to a dangerously low level.

Control of Floating Plants

The most common floating plants are duckweed, water lily, spatterdock, and watershield. Water lily plants are valued by fisherman as long as ample room is left for

FIGURE 4. Biological control of alligatorweed by the alligatorweed flea beetle, *Agasicles hygrohila*. Observations being taken by Dr. N. R. Spencer, entomologist. (Photograph courtesy of the U.S. Department of Agriculture, Washington, D.C.)

FIGURE 5. Atlantic Intercoastal Waterway near Norfolk, Va., maintained by the U.S. Army Corps of Engineers. Weeds controlled by chemical methods. (Photograph courtesy of the U.S. Army Corps of Engineers, Norfolk district.)

FIGURE 6. St. Johns River near Aster, Fla., maintained by the U.S. Army Corps of Engineers. Weeds controlled by chemical methods. (Photograph courtesy of the U.S. Army Corps of Engineers, Jacksonville district.)

fishing, but in areas used for boating, swimming, or water skiing, water lilies are a serious nuisance and must be kept under control.

Plants and the water surface should be treated by spraying, taking care to obtain even distribution in the area of plant growth. If water is colder than 60°F, or if plants are not growing actively, the area should not be sprayed. For floating plants that are rooted on the bottom, granular formulations may be more effective.

Control of Submersed Plants

Submersed plants complete their entire life cycle, exclusive of flowering and pollination, under water. Waterweed is typical of submersed plants that grow in dense stands or mats. Other troublesome submersed species that grow densely include, pondweed, water milfoil, coontail, fanwort, and bladderwort.

Chemicals for controlling submersed weeds can be applied in various ways. Granules may be spread from a boat, from the shore, or by aircraft. Care must be taken to obtain uniform coverage. Granules are especially useful for spot treatments near boat landings or swimming areas. Liquid herbicides may be sprayed or pumped into the water. Again, boats, aircraft, or even hand pumping from the shoreline may be used. As a rule, the more evenly the chemical is distributed, the better the results will be. Spot treatment of problem areas with liquids in very large areas may be desirable. Usually more material per acre-foot of water is needed when making spot treatments than when making overall treatment because the nontreated adjacent water will have a diluting effect. Where dense weeds cover an entire pond, only one third of the pond should be treated at a time, to prevent oxygen depletion due to decaying plant growth. When treating one third of the pond, applications should be made at weekly intervals until the entire ponds is treated.

Control of Emersed Plants

Emersed plants may grow in moist, swampy shoreline soils and extend into water, 3 ft or more in depth. Some emersed plants are valuable duck foods and should not be eliminated if waterfowl are wanted. Principal types include species of rushes, bulrushes, cattails, reeds, sedges, and arrowheads.

Foliar sprays should be applied in early summer as the leaves are approaching full size, but before flowering. If water is used as the diluent or carrier, a small amount of detergent or other wetting agent will improve coverage of the plant surfaces. The plant foliage should be completely wetted. Applications should be made during midday periods when the foliage above water is free from dew or rain.

Granular formulation act more slowly than liquids, but they may give more long-lasting results. If attaclay or a similar carrier is used, the herbicide may be spread with a cyclone seeder or by hand. The granules settle to the bottom and release the chemical that will be absorbed through the plant roots. As with foliage sprays, granular applications are made in late summer or fall, however, the effects may not be noticed until the following spring, when the treated plants fail to to develop. If granules are spread on choppy water, distribution will be uneven, and results may be unsatisfactory.

In making spray applications for controlling aquatic weeds, there are two basic methods of calculating the dosage and application rate: (1) quantity of herbicide per surface area treated and (2) quantity of herbicide per volume of spray solution.

For treating areas 1 acre or more, the first method should be used. The important consideration with this method is the amount of herbicide applied per acre. The amount of carrier will vary with the spray equipment, nozzle size, and pressure. For treating areas of less than 1 acre, it is more practical to use the herbicide per volume method and to spray foliage to the run-off point or until thoroughly wetted. With most herbicides, the container label will give directions for applications by both methods.

Conditions of Application

The applicator should consider only those herbicide formulations specifically registered and labeled for aquatic weed control.

The 3- to 5-gal pressure-type sprayers available at most home garden centers and hardware and farm supply stores are adequate for treating small ponds. Ponds of 5 acres or more may be more efficiently treated with heavier, commercial-type spray equipment.

Granules can be spread by hand in small areas such as boat landings and swimming sites. Hand-operated, crank-type fertilizer and seed spreaders will do a satisfactory job in larger areas. Large versions of these spreaders are used for large-scale commercial granule applications. They operate on the principle of the cinder spreaders used on highways and are capable of dispersing granules in 30- to 35-ft swaths. These units are often powered by 12-V DC motors that operate from an auto battery. Helicopters and airplanes have been used effectively to apply chemicals, particularly granules. In large-scale operations, chemicals, especially silvex sprays, should be applied on a calm day when dangers from drifting can be avoided. With such precautions, damage to desirable vegetation is minimal, and operators are protected from excessive contact and inhalation of spray.

Chemical spillage, especially around stock pond areas or where pets and children may be exposed, must be cleaned up. Herbicides should never be stored where children can reach them. Labels should be studied and their instructions for safe and proper use observed.

Herbicide application can start at the margins of the lake or pond and progress toward the deep water. Fish and other organisms will tend to move ahead of the herbicide-treated water into the deeper areas where toxic concentrations can be avoided.

Lakes and ponds that have a high rate of inflow should be lowered prior to treatment so as to ensure adequate contact time for the herbicide. The spillway should be closed, and the treatment water should be retained for at least 3 days before overflow. The amount of drawdown will vary according to the situation. Care must be taken to avoid excessive exposure of bottom areas.

After the application is completed, the spray equipment should be flushed with large quantities of water, and the wash water should be properly disposed to prevent contamination of water resources.

Water areas should be treated for aquatic weeds in late spring or early summer when plants are young and actively growing. Treatment at this time of year is usually given optimal control with a minimum of chemical. Applications in late summer or early full require more chemical and usually give a slower, erratic control. Furthermore, access to normally dense aquatic vegetation is best in the spring or early summer before the plants reach the surface.

The applicator should:

1. Be certain of the identity of the vegetation to be treated and the capability of the chemical to control it.
2. Obtain a permit to treat any lake or pond, whether for weed control, cleaning, or other purposes.
3. Never use more chemical than suggested on the label. Fish and other valuable aquatic life may be killed.
4. Use treated water only as suggested on the label.

Environmental Changes

The sudden elimination of a dense growth of vegetation from an aquatic environment very often causes side effects that can produce significant changes in the biological and physical makeup of a lake, pond, or stream.

Following the death of larger weed plants in a pond or lake, a greenish or yellowish-brown turbid condition may be noticed. This condition is due to the presence of billions of microscopic algal cells which have utilized the nutrient for growth and reproduction. The blue-green algae are often responsible for a green pea-soup appearance, whereas other algae and various one-celled organisms cause the yellowish-brown colors in water. When conditions are optimal for development of algae, a dense bloom can develop quickly. These dense blooms of plankton algae cut down light penetration and thereby inhibit the reestablishment of those species killed in treatment, but these algae may be more objectionable than those killed in treatment.

Some aquatic vegetation is necessary for the reproduction and survival of certain fish (pickerel, golden shiners, and others). It follows that where desirable fish are dependent on aquatic vegetation, portions of water should be left untreated.

Aquatic plants provide habitat in lakes and ponds for waterfowl and fish, but may be in conflict with certain types of water use.

Aquatic plants add essential oxygen to the water, but under some conditions, their respiration and decomposition may reduce the oxygen to dangerously low levels for fish survival.

Aquatic plants perform an important role in food chains; however, too many plants may contribute to overcrowded and poor-quality fish populations.

There is no easy cure-all for controlling undesirable aquatic vegetation. Dredging to deepen shallow edges may help, but may also create hazards and fish management problems.

Aquatic plant control with herbicides can be successful and very satisfactory. Once a weed problem is under control, diligent treatment of regrowth is necessary to maintain control. It is much easier and less expensive to conduct periodic maintenance or to make spot treatments than it is to wait until treatment of the entire area is necessary.

ACTIVITIES AFFECTED

Throughout the conterminous U.S. and Puerto Rico, and particularly in certain portions of the southeastern U.S., the obnoxious growth of aquatic plants adversely affects every kind of water-based activity. The following paragraphs discuss effects on navigation, recreation, flood control, drainage, agriculture, public health, and fish and wildlife conservation.

Navigation and Power Generation

Aquatic plants have disrupted navigation, mainly in the southeastern U.S. Nationwide, the navigable waterways support a traffic load of over 1 billion ton/year. It is estimated that 15 to 20% of these waterways are susceptible to infestation by aquatic plants. In addition to directly disrupting commercial and recreational navigation on these waterways, aquatic plants have indirect effects. Floating plants collect in low flow and eddy areas around locks, flow-control structures, and navigation devices. Aquatic plants have often reduced the operational efficiency of these structures. Submersed weeds have the same effects by temporarily fouling underwater moving mechanisms of navigation control structures as well as impeding boat traffic.

Recreation

Large growths of emersed weeds deny access to swimmers, boaters, and fisherman. Prolonged existence of certain of these species can destroy production of fish food sources due to the absence of light beneath the surface. Once the plants are out of control, their massive destruction requires intensive in-place control measures which result in sinking mats of organic materials. This large organic load reduces the dissolved oxygen in the water, often resulting in fish kills. There are over 29,500,000 acres of fishable freshwaters in the U.S. It is estimated that people spend 300,000,000 activity days per year in freshwater recreational fishing at an estimated value of approximately $1.50 to $2.00 per activity day. Commercial freshwater fishing accounts for more than $400,000,000/year. Fishing, however, is not the only recreational activity affected by obnoxious growths of aquatic plants. Water skiing is a multimillion dollar industry affected by aquatic plant growths. Swimming is curtailed or prevented in many lakes where emergent or submergent plants proliferate. For example, a death in Florida has been attributed to the submersed weed, hydrilla. Similarly two deaths in Oklahoma were attributed to entangelement of swimmers with waterweed (*Elodea*).

Flood Control

It is estimated that 75% of the total annual runoff takes place over a relatively short period of time and, in certain topographic contexts, has caused considerable damage through flooding. Storage reservoirs provide buffers in these drainage systems so that controlled releases of the water volumes prevent downstream flooding as a part of overall water-management practices in a drainage basin. These reservoirs provide an increased water surface in which aquatic plants grow. Uncontrolled, these plants can inhibit the free flow of water through control structures as well as provide a source of plants for downstream infestations subsequent to reservoir releases.

Drainage, Agriculture, and Public Health

Aquatic plants have clogged drainage canals to the extent that small-scale, short-term flooding has resulted. These plants have also clogged agricultural drainage canals to the extent that agricultural lands cease to drain efficiently. In humid areas, less than half of the runoff can be depended upon for irrigation use, and in arid and semiarid areas, only about 25% can be considered a dependable supply. Storage reservoirs in these areas are critical for the prevention of severe water shortages. Runoff often flushes plants from natural wetlands into these reservoirs. As land-based peripheral activities increase the nutrient inflows to these reservoirs, more and more of these facilities are providing ideal growing conditions for aquatic plants. Once there is a significance coverage with aquatic plants, evapotranspiration increases the water lost to the atmosphere. In the case of multipurpose reservoirs, the control required as a result of the presence of these plants can affect the quality of the water.

Storage facilities used for water-supply systems are not very amenable to repetitive treatment with chemicals. With an increased requirement of the water, public health can be affected. Emergent floating aquatic plants provide breeding habitats for mosquitoes. These pests, protected from control measures by the plants themselves, are disease carriers and therefore a potential detriment to public health.

Fish and Wildlife Conservation

There are approximately 169,927,000 acres (109,206 mi^2) of wetlands which include swamps, marshes, and bogs. These areas are natural spawning grounds for aquatic plants. During natural water fluctuation, the aquatic plants find their way into the waterways, streams, and lakes. To treat these aquatic plants at their source is obviously attacking the heart of the problem. However, these same areas abound with wildlife spawning grounds. For this reason, these areas are not amenable to repeated treatment by chemical control methods. In addition, the fish and wildlife management practices have been affected by aquatics in water bodies simply because their presence may deny access of fish and waterfowl to water surface areas.

Inland waters on which these activities take place cover a surface area of 54,174 mi^2. This total reflects those lakes, reservoirs, and ponds larger than 40 acres in area (Great Lakes excluded) and streams and canals one eighth of a statute mile or more in width. Included in this total area there are over 29,500 mi of navigable channels authorized for improvement by the Congress of the U.S. under the Rivers and Harbors Acts. At the present, the Corps of Engineers is maintaining over 25,000 mi of the inland waterways.

PROBLEM SPECIES

The water-covered surfaces in the U.S. comprise slightly over 2% of the total surface area; the social, environmental, and economic value to the nation is enormous when consideration is given to navigation, flood control, drainage, etc. This vast water-covered area of over 105,590,000 acres (164,980 mi^2) is the potential habitat (depending on the suitability of growing conditions in a specific locality) for more than 50 different aquatic plants which have become problems of various degrees depending on the type and intensity of water use. These species presently identified as problem aquatic weeds are listed in Table 1 according to the problem site and species.

Problem Site and Species

Presently, the most prevalent problem species occurring in high-use areas are for exotic species: waterhyacinth (*Eichhornia crassipes* (Mart.) Solms), alligatorweed (*Alternanthera philoxeroides* (Mart.) Griesb.), hydrilla (*Hydrilla verticillata* Royale), and waterlettuce (*Pristia stratioties* L.). In addition to these four, three other species have presented continuing problems: Eurasian watermilfoil (*Myriophyllum spicatum* L.), *Egeria densa* Planch., and waterchestnut (*Trapa natans* L.). Salt cedar (*Tamarix* spp.) is a serious phreatophyte problem in the southwest.

Table 1
AQUATIC PROBLEM PLANTS, ESTIMATED INFESTED ACREAGE, TOTAL SURFACE ACREAGE, AND THE POTENTIAL PROBLEM ACREAGE[9]

Aquatic problem plant	Estimated infested area (acres)	Total surface area available (acres)	Potential problem area (acres)
Floating			
Waterhyacinth	1,000,000	38,200,000	9,550,000
Waterlettuce	3,000	38,200,000	9,550,000
Waterchestnut	3,000	3,000,000	1,050,000
Waterfern	3,000	3,000,000	1,050,000
Submersed			
Hydrilla	50,000	12,300,000	4,305,000
Egeria	50,000	12,300,000	4,305,000
Watermilfoil	250,000	69,800,000	17,450,000
Potamogeton	1,000,000	69,800,000	17,450,000
Emersed			
Alligatorweed	60,000	39,400,000	9,850,000
Other rooted-emersed	1,500,000	69,800,000	17,450,000

Considering the rapidity with which most aquatic weeds spread, it is emphasized that the concern should be on the magnitude of the potential problem as well as on the presently infected water bodies. The potential problem acreage for the geographical areas within the boundaries are given in Table 1.

As shown in the tabulation, not every available acre of water surface is a potential problem acre. Water quality, environmental conditions, seasonal cycles, and plant competition all play a role in whether or not a particular plant will be able to flourish in a given locale. The estimate of potential problem acreage is based on a percentage of the total available water surface within the bounded region of known infestation for a particular plant species. The estimated percentage varies from 10 to 85, depending upon the plant species, the particular site and the known or estimated environmental conditions.

Project Effectiveness

Considering present control techniques and the prolific growth rate of some of the plants, the total acreage could increase by an amount equal to five times the acres being controlled. This would result in a net total of over 1,000,000 acres of waterhyacinth in 2 to 3 years. For example, the waterhyacinth can, under ideal conditions, double in numbers every 12 to 14 days.

The present arsenal of control methods available can be grouped into the following categories:

1. Chemical — The use of herbicides by any application means to effect rapid, but temporary, control.
2. Mechanical — The use of mechanical devices or machines to physically alter or remove the plants from the water body.
3. Biological — The use of biological agents, such as insects, pathogens, and fish, to effect a slower, but more permanent, control.
4. Integrated — The use of chemical, mechanical, and biological means in combinations such that effective permanent control can be achieved.

5. Environmental manipulation — Sequential management of selected life-cycle events for the control of target species, without the introduction of foreign agents or organisms into the ecosystem.

To date, chemicals have been by far the most widely used technique for control on an operational scale. This is understandable when consideration is given to the fact that immediate solutions are demanded by the public, and chemicals presently provide the only predictable and timely response for almost all of the problem species. In recent years, biological control methods have been used successfully for some problem species, e.g., alligatorweed. This method often takes several years to reach a level of control demanded by the public, and the effectiveness is often unpredictable. Integrated control techniques show promise in permitting management decision trade-offs that properly take into account predictability, control effectiveness, costs, and environmental compatability. However, with no exception, each of the present control methods is inadequate for the control of all the aquatic pest plants.

COST AND POTENTIAL SAVINGS

Typical costs for current control operations using herbicides and mechanical techniques are about \$30 and \$800/acre, respectively. General cost histories for the biological and integrated control techniques are not available for most aquatic weeds. However, an example of costs savings for biological control can be obtained by reviewing the use of the alligatorweed flea beetle to bring alligatorweed under control. In this successful program, approximately \$1,000,000 (including research costs) was expended over a 10-year period and control has been effective for over 4 years. This has resulted in an estimated savings of \$20/acre for treatment costs by relieving the necessity to use herbicides. Prior to introduction of the beetle, chemicals were used for alligatorweed control, and there were an estimated 160,000 acres of this weed. The cost of bringing the 160,000 acres under control by chemical means would have been \$3,200,000. Because herbicides control the weed only temporarily, reapplication to the same areas would have been necessary four times during the 10-year period at a total cost of \$11.2 million. The benefits provided to the operational costs by the research program exceed the costs by a ratio of 11:1. For the Corps of Engineers research program alone, which funded a major portion of the alligatorweed flea beetle program each year, the research benefit-cost ratio is estimated to be 8:1 in terms of savings to the operational program.

Also, there is impressive potential for cost saving through the use of biological controls of waterhyacinth. For example, aquatic areas of infestation of the southeastern U.S. were reduced by 35.5% during the 4-year period from 1959 to 1963, primarily through use of chemicals. Yet, at the end of the next 10 years (by 1973), the acreage of waterhyacinth in this same 8-state area was estimated to be over 300,000. During this 14-year period, an annual average of 35,000 acres, or a total of approximately 525,000 acres of waterhyacinth as under treatment.

Recent research in biological control of waterhyacinth indicates that the approach is feasible and control can be obtained at significant savings over chemical treatment.

No data are available to calculate the cost savings that would result from implementing integrated control methods; nevertheless, there is a recognized need for integrated control techniques on an operational scale using chemical (for quick stresses) and biological agents (for maintenance and long-term control).

One of the major benefits of improved control techniques through operationally oriented integrated management is environmental compatibility. The management tools necessary to respond to the public requests for pest plant control, must be obtained without violating the National Environmental Protection Agency (NEPA) regulations. Although control cannot now be accomplished in all situations cited, the cost-effectiveness of plant control will bring public pressure on the need for control. For example, House Document No. 251, 89th Congress, 1st Session, Expanded Project for Aquatic Plant Control, July 28, 1975, indicated that a national average of benefits derived from operational control programs would exceed the costs by from 5:1 to 10:1. Data compiled in the Office of the Chief of Engineers in 1969 estimated that as a result of $6,000,000 federal and $2,000,000 nonfederal funds expended over the 8-year period, 1959 to 1967, benefits to water supply, fish and wildlife, recreation, and public health exceeded costs by 4:1. Present estimates based on escalating costs and increased recreational activities and developments place the ratio at approximately 6:1. Using the more conservative figure and an estimated $21,000,000 being presently spent annually by state and federal agencies, one obtains an annual estimated savings of $216,000,000 as a result of aquatic weed control operations.

REFERENCES

1. Department of the Army, Herbicide Manual for Noncropland Weeds, Technical Manual TM 5-629, U.S. Government Printing Office. Washington, D.C., 1970.
2. Department of the Navy, Basic Pest Control Technology, Naval Facilities Engineering Command, Alexandria, Va., 1976.
3. Department of the Navy, Chemical Control of Disease Vectors and Economic Pests, Naval Air Station, Jacksonville, Fla., 1976.
4. **Stewart, L. E., Gienger, G. W., Parochetti, J. V., and Coan, R. M.,** Broadcast Spraying, Extension Bulletin 229, University of Maryland, College Park, 1976.
5. **Bartley, T. R. and Gangstad, E. O.,** Environmental aspects of aquatic plant control, *J. Irrig. Drain.*, 100, 231, 1974.
6. **Frank, P. A., Hodgson, R. H., and Comes, R. D.,** Evaluation of herbicides applied to soil for control of aquatic weeds in irrigation canals, *Weeds*, 11, 124, 1963.
7. **Gangstad, E. O.,** Aquatic plant control program, *Hyacinth Control J.*, 9, 46, 1971.
8. Research Planning Conference on Integrated Systems of Aquatic Plant Control, Proceedings, October 29 to 30, 1973, U.S. Army Waterways Experiment Station, Vicksburg, Miss., 1974.
9. Report of the USDA Research Planning Conference on Aquatic Weed Control, Department of Botany, University of California, Davis, U.S. Army Corps of Engineers, Washington, D.C., 1977.

Chapter 18

MANAGEMENT OF PHREATOPHYTE AND RIPARIAN VEGETATION

INTRODUCTION

The most undesireable invading wood perennial phreatophyte in the southwest is saltcedar (*Tamarix pentandra*). During the past three decades it has spread over most of the southwest. It has no agronomic value and is a serious threat to the economy of that area because it

1. Displaces native range plants
2. Causes a serious consumptive loss of ground water
3. Is a potential flood hazard,
4. Is a serious fire hazard

Control of saltcedar is largely limited to mechanical methods, but in areas where this is not possible, herbicides must be used. Occasionally applications of Silvex or frequent applications of 2,4-dichlorophenoxy (2,4-D) at the early leaf stage can be used for control. Other riparian species, such as alder, cottonwood, persimmon, sage brush, willow, and yucca, are best controlled with Silvex®, but can be controlled with dicamba or dicamba plus 2,4-D.

PECOS RIVER BASIN WATER SALVAGE PROJECT

This project was authorized to:

1. Prevent further decreases in the supply of water in the Pecos River Basin
2. Enhance the water supply for municipal, industrial, irrigation, and recreational uses
3. Provide protection for farmlands in the basin from hazards of floods
4. Provide for the conservation of fish and wildlife

These purposes will be accomplished by a management program to reduce the growth of saltcedar and other phreatophytes.

Between Santa Rosa, N.M. and Girvin, Tex. are approximately 590 mi of the Pecos River and, before clearing, over 200,000 acres of saltcedar (*Tamarix* spp.). Lateral boundaries of the area previously cleared were established on each side of the river to a point where depth of the water table was no more than 10 ft—a relatively high water table providing suitable growing conditions for phreatophytes. Within these confines were approximately 78,000 acres of lands, most of which were vegetated with dense stands of saltcedar (see Figures 1 and 2).

Clearing activities conducted by the Bureau of Reclamation began in 1967 and continued until 1971. About 53,950 acres at various locations between Lake Summer, N.M. and Pecos, Tex., a distance of about 370 mi, were within the clearing boundaries. No new clearing has been done since 1971. Maintenance activities on 53,800 acres of cleared areas were initiated in 1968 and are continuing.

Hydrologically and environmentally oriented studies are underway and pending to further evaluate project effects on the broad-based, interrelated environmental and economic resources of the Pecos River Basin. Upon completion of the studies and analyses, the Pecos River Basin Water Salvage Project will be reevaluated, and should

FIGURE 1. Saltcedar (*Tamarix*) spp. invasion of west Texas rangeland along the Pecos River near Girvin, Tex. (Courtesy of the Texas Agricultural Experimental Station, College Station.)

the evaluation indicate a change from a program of continued maintenance, further controls will be initiated.

Congressional hearings on Senate Joint Resolution 9, May 1963, and subsequent reports and amendments from the Senate (Report No. 572, 88th Congress, 1st Session) and the House of Representatives (Report No. 1572, 88th Congress, 2nd Session) were the basis for project authorization. Public law 88-594, dated September 12, 1964, authorized the Secretary of the Interior, as he deems necessary, to carry out a continuing program to reduce nonbeneficial consumptive use of water in the basin including that by saltcedar and other phreatophytes.

Phreatophytes are defined as plants that habitually grow where they can send their roots down to the water table or capillary fringe area immediately overlying the water table and are able to obtain a perennial and secure supply of water. The word "phreatophyte" is derived from two Greek roots—"phreatos" (a well) and "phyte" (a plant with a particular characteristic of habit—and thus means "well plant". In this statement, the term "phreatophyte" is confined to those nonagricultural plants that do not have significant commercial value. Phreatophytes normally grow where the ground water table is relatively high (usually less than 10 ft below the soil surface); they are commonly found growing along streams, reservoirs, or irrigation canals.

The major phreatophyte species which is a problem in the southwest is the saltcedar. Saltcedar was first noted in the McMillan Delta area, Pecos River Basin, in 1912. Since that time saltcedar has spread throughout the basin, occupying more than 275,000 acres in 1961. Saltcedars contribute to the problems of water shortages and flooding within the Pecos River Basin area. It is believed that the annual rate of water consumption by saltcedar is greater than that of any other phreatophyte and that saltcedar and other phreatophytes consumptively use a large amount of water. In addition to contributing to the water shortage, saltcedars aggravate flooding during periods of high

FIGURE 2. Closeup of the dense stand of saltcedar; very difficult to control by most methods, but best controlled by spraying with Silvex®. (Courtesy of the Bureau of Reclamation, Amarillo, Tex.)

riverflow by invading and restricting normal stream floodways; reducing water-carrying capacity of the floodway; choking stream channels; impeding water flow; and forcing high flows to disperse over a larger area. Phreatophyte invasion onto farmland can also decrease crop yield and increase costs associated with weed control.

The water supply from the Pecos River drainage basin and from Avalon Reservoir, Lake McMillan, and Lake Summer for the Carlsbad Project, New Mexico, has recently been meager. In addition, the waters of the Pecos River Basin are overappropriated. The Carlsbad Irrigation District (CID) has water rights, adjudicated by the 1933 Hope Decree, with priority dates earlier than most other water users in the basin. Lack of water for the project has prompted the CID to request that the New Mexico State Engineer administer water of the Pecos pursuant to the Doctrine of Prior Appropriation.

PREVIOUS CLEARING METHODS AND ACTIVITIES

Several methods involving mechanical and chemical means have been used to clear phreatophytes and maintain cleared areas. These methods will be discussed under the headings (1) mechanical methods, (2) chemical methods, and (3) mechanical and chemical methods.

Nonchemical Methods

Various mechanical methods which have been used to clear wood phreatophytes include tree crushing, disk harrowing, bulldozing, root plowing, chaining, brush raking, chopping, rotary mowing, and draglining. The preferred method of initial clearing consisted of a combination of chaining followed by root plowing to smooth the area. Operation and maintenance programs usually consisted of annual root plowing or grubbing depending on the amount of regrowth. In areas with a high percent of saltcedar regrowth, root plowing was preferred; in areas with a low percent of saltcedar regrowth, grubbing was preferred.

Depending on the circumstances, the mechanical methods listed below were used; therefore, a description of each of these methods is included even though they may not have been the preferred method.

Chaining— A chain, 360 to 400 ft long, weighing about 40 to 50 lb/ft, has been used for clearing. The chain is doubled (folded in half) and pulled between two crawler tractors and can be weighted in the middle for added efficiency. Chaining either uproots plants or shears trunks near ground level. Chaining worked best in areas with little underbrush and where tree trunks are 4 to 18 in. in diameter. Chaining was also used to smooth the ground after root plowing, knock down heavily weeded areas, and breakup dead debris left from previous operations.

Root plowing—Root plowing shears vegetation below the soil surface. The root plow is designed for killing medium-to-large phreatophyte brush by undercutting vegetation below the root crown or bud ring. Root plowing was effective on vegetation where density is nearly 100%.

Grubbing—Grubbing is similar to root plowing; however, the grubber blade is much smaller than the root plow. Grubbing was used in clearing small stands or clumps of phreatophytes and was not as destructive to the habitat as root plowing.

Chopping—Mechanical chopping knocks down and shatters trees 6 to 7 ft in height and 2 to 4 in. in trunk diameter. The drum of the chopper was welded-on cutting blades which can penetrate the soil to a depth of 6 to 12 in. These blades cut, fracture, and shatter growth, leaving the woody vegetation in a flattened mat. Fallen trees up to 12 to 15 ft in length and 6 to 8 in. in trunk diameter can be chopped.

Tree crusher—Tree crushing flattens and crushes vegetation up to 15 inches in trunk diameter. The tree crusher breaks the trunk of the saltcedar near ground level and cuts or crushes the entire tree into small pieces. This method of clearing was used extensively in the past, but is not expected to be used in the maintaining program.

Disk harrowing—Disk harrowing crush-pulverizes the ground surface and uproots and cuts small-sized vegetation. It was not suitable for removing strong woody plants.

Bulldozing—Bulldozing shears plants near ground level and uproots whole plants.

Brush raking—Brush raking shears small brush and stumps below ground level and pushes them into piles. Rakes have the advantage of permitting the soil to pass between the teeth as they are pushed through the soil, ripping out and pushing brush, stumps, rocks, and other debris. Rakes were most effective in the sandy soil of alluvial flood plains in pushing sheared, uprooted, or broken woody vegetation into piles or windrows.

Rotary mowing—Rotary mowing clears woody vegetation up to about 2 in. in diameter. Mowing was most effective in removing above-ground vegetation left by heavier clearing equipment and for maintaining areas previously cleared of phreatophytes. It severs the vegetation about 10 in. above ground level and shreds it into small pieces.

Draglining—Draglining was used to clear vegetative plugs and incidental vegetation from the channel. It was not generally suitable for large vegetation and flat area clearing.

Chemical methods—Herbicides which have been used to control phreatophytes include 2,4,5-TP (Silvex®), 4(2,4-dichlorophenoxy) butyric acid (2,4-DB), and 2,4-D. Of these, Silvex® is the herbicide presently registered by Environmental Protection

Agency (EPA) for use on saltcedar. Herbicides were applied at a rate of 3 lb active ingredient (ai) per acre. The dilution rate for application is 3 lb ai in 8 to 20 gal of water. The following criteria were used for the application of herbicides.

1. In areas where herbicidal drift could be a problem, an invert emulsion was used to reduce drift.
2. No ground spraying when wind velocity reaches 8 mi/hr.
3. No spraying within ½ mi of susceptible crops.
4. No spraying within 100 ft of any navigable water.
5. Monitoring of streams during spraying season.
6. Strict adherence to recommendations and instructions which apply to the herbicide used.

The herbicides which have been used are growth hormones and are most effective when food (carbohydrate) reserves are at their lowest levels. The carbohydrate reserves of saltcedar plants are at the lowest level during the months of April, May, and June.

The Bureau of Reclamation has taken steps to observe the effect that herbicide use within the flood plain of the Pecos River might have on downstream water quality. A cooperative herbicide monitoring program between the Bureau of Reclamation and the Geological Survey was initiated in 1971. The monitoring program has been carried out each year that aerial spraying methods were used.

Mechanical and Chemical Methods

In past clearing operations, the chem-cut rotary mower was utilized. This method consists of a standard rotary mower that has been modified by adding spray nozzles for herbicides. The spray nozzles and cutting blades are covered by a protective metal housing. The herbicide (Silvex®) is usually applied as an invert emulsion on the shattered stumps of wood growth. This treatment method is not currently planned for use in future control operations since it was not as effective as other methods.

Previous Clearing Patterns

Clearing of saltcedar was restricted to areas with an areal density (tree crown density) greater than 10% and having a depth to the water table of 10 ft or less. Trees and brush were uprooted or cut level with the average ground surface. In areas infested predominately by growth small enough to allow mowing, vegetation was controlled by mowing at a height of not more than 6 in. (4 in. in the Bitter Lake Wildlife are) above the average ground surface.

COST EFFECTIVENESS OF THE CONTROL PROGRAM

Although other mechanical methods have been used, mechanical maintenance activities in 1975, 1976, and 1977 have been primarily root plowing and grubbing (see Table 1). Other mechanical means may be used in the future, but root plowing and grubbing are more selective and achieve better saltcedar control. Therefore, it is anticipated that root plowing and grubbing will be the primary methods used in the future to control saltcedar on previously cleared areas.

Maintenance requirements may change annually depending on regrowth. Determination of the methodology to be used and the specific areas to be maintained are based on the experience of the manager. Management practices anticipated for the near future are for mowing/herbicide application, 2000 acres; root plowing, 4000 acres; and grubbing, 48,000 acres. It should be noted that the above acreage are gross acres within clearing boundaries and do not necessarily mean that every acre within the boundaries

Table 1
METHODS USED TO CONTROL PHREATOPHYTE REGROWTH AND THE ACRES CONTROLLED BY EACH METHOD SINCE 1968

Method	1968	1969	1970	1971	1972	1973	1974	1975	1976	1977
Herbicides	2,000	1,985	6,145	8,900	9,914	9,057	7,964	1,980[a]	1,879[a]	2,013
Chopping	2,500	11,750	26,961	26,175	20,940	15,025	3,790	—	—	480
Chem-cut	—	60	30	474	—	—	—	—	—	—
Mowing	—	1,715	3,944	4,213	2,373	3,925	205	70	—	190
Root plowing	—	—	—	—	5,415	19,565	16,460	21,235[a]	11,698[a]	5,844
Grubbing	—	—	—	703	2,608	—	15,000	36,174[a]	51,805[a]	50,411
Chaining	—	—	—	270	100	4,250	5,850	15,811	16,727	8,200
Total	4,500	15,510	37,080	40,735	41,350	51,822	49,269	75,270	82,109	67,138

[a] Some areas have undergone maintenance activities two or three times in a single year.

From U.S. Department of the Interior, 1979 Pecos River Basin Water Salvage Project, New Mexico-Texas, Final Environmental Statement, FFS 79-9, Southwest Region, Amarillo, Tex., 1979.

will be retreated. Grubbing, for instance, is primarily associated with individual plants. The mowing/herbicide application technique provides for mowing the area one year and applying herbicide the next year. This technique was put into practice in 1977.

Table 1 indicates that herbicides have been applied annually on 2000 to 10,000 acres of land. Herbicides were applied on 1980 acres in 1975, 1879 acres in 1976, and 2013 acres in 1977. No herbicides were applied in 1978. The total area treated annually varied depending on regrowth conditions and location.

INTERRELATIONSHIP WITH OTHER PROJECTS OR PROPOSALS

The Pecos River Basin Water Salvage Project is interrelated with the following projects.

Los Esteros Flood Control Project

This Corps of Engineers project was authorized by the 1954 Flood Control Act, Public Law 780, 83rd Congress, and is currently under construction. This project will involve conversion of a large part of the irrigation storage capacity of existing Lake Summer to flood-control use and a transfer of the converted irrigation storage space upstream to an earthfill dam to be constructed at the Los Esteros site. The Los Esteros Project will be for flood control, sediment storage irrigation, and possibly for municipal water supply. The Los Esteros site is located about 7 mi north of Santa Rosa, N.M. on the Pecos River. Reservoir storage capacity at the spillway crest will be 449,000 acre-ft.

Other Projects

Corps of Engineers work includes several completed flood-control projects in the basin. The largest project is the Two Rivers Reservoir, completed in August 1963, to protect the city of Roswell and other downstream areas from floods originating on the Rio Hondo and Rocky Arroyo. The dam is located a few miles west of Roswell across both of these tributaries of the Pecos River project in New Mexico is the floodway and diversion works on Dark Canyon.

Fort Sumner Project

The Fort Sumner Project is a private irrigation project reconstructed and rehabilitated by the Bureau of Reclamation. The irrigable area of 6500 acres that can receive a full supply of water from the project is northeast of the Pecos River immediately below the town of Fort Sumner, New Mexico. Principal structures are the Fort Sumner Diversion Dam, pumping plant, the main and Highline Canals, and several large drains.

McMillan Delta Project

This is an authorized Bureau of Reclamation project which, when constructed, will provide a 16-mi (1500 ft^3/sec) conveyance and drainage channel with a cleared floodway through the delta. The delta is an almost flat plain varying in width from 1.5 to 3 mi. There are about 13,500 acres in the delta, consisting of over 10,000 acres of dense saltcedars and varying amounts of different grasses.

Public Law 85-333, dated February 20, 1958, authorized the McMillan Delta Project, but prohibited clearing of the floodway until provisions had been made to replace any Carlsbad Irrigation District terminal storage which might be lost by the clearing. This project is summarized in House Document 429, 84th Congress. Brantley reservoir (described below), when constructed, will provide the necessary replacement of terminal storage and, therefore, eliminate restrictions for additional clearing of phreatophytes. It is estimated that there are about 10,000 acres which would require selective clearing in McMillan Delta.

Brantley Dam and Reservoir

The Brantley Project is an authorized Bureau of Reclamation project. Authorization was obtained in 1972 under Public Law 92-514. The project would consist of a dam and reservoir located on the Pecos River in Eddy County, N.M., about 14 mi upstream from Carlsbad. Final design of the dam is not complete at this time. A final environmental statement of this project (FES-72-30) was filed with the Council on Environmental Quality on September 11, 1972.

Carlsbad Project

The Carlsbad Project is in southeastern New Mexico in the vicinity of Carlsbad. The privately owned project was rehabilitated by the Bureau of Reclamation in 1906. Project features include Lake McMillan and Avalon Reservoir, which were reconstructed in 1906, and Lake Sumner (formerly Almogordo Reservoir), which was built in 1937. Project features also include drainage and distribution systems to irrigate 25,055 acres of land in the vicinity of Carlsbad, N.M. Lake Sumner provides flood control and recreational benefits although there is no storage allocated for these purposes. The irrigation plan for the project provides for storage of water in Lake Sumner, Lake McMillan, and Avalon Reservoir and the diversion of the water from Avalon Reservoir into a canal system irrigating about 25,055 acres of land on both sides of the Pecos River near Carlsbad, N.M. The operation and maintenance of the project were assumed by the Carlsbad irrigation District in 1949, except for Sumner Dam and Lake Sumner which remain the responsibility of the Bureau of Reclamation.

Red Bluff Water Power Control District

The Red Bluff Water Power Control District is privately owned and is the main district which administers the release of water for seven downstream irrigation districts. It consists of Red Bluff Dam and Reservoir. The seven downstream districts operate the numerous downstream diversion dams.

Watershed Flood Protection

Under Public law 83-566, the Soil Conservation Service has engineered and constructed a large number of flood-control and watershed-protection projects in the Pecos River Basin in a cooperative program with the New Mexico State Engineer Office and local agencies. The first projects were initiated in 1955 and are designed to provide protection for watersheds of less than 250,000 acres. Since 1955, numerous stabilization and sediment-control structures and debris basins have been constructed, several miles of stream channel improvement and diversion dikes and ditches have been completed, and thousands of acres in the basin have been subjected to treatments, such as terracing and contour plowing. Completed projects are the Upper Rio Penasco, Zuber Draw, Pecos Arroyo, and Hackberry Draw.

INTEGRATED MANAGEMENT

The optimum management of the alluvial flood plains, overgrown with saltcedar and other riparian vegetation, must take account of the various environmental conditions that pertain to a particular use site.

In the very shallow water table areas (0 to 4 ft), grazing and flood passageways are best suited to this area, and efforts should be made to establish and maintain a grass cover. Chemical control is best suited for this area, or shrubs can be cut with a rotary mower or brush hog.

For areas when the water table is down 4 to 8 ft, saltcedar is desirable for wildlife purposes, but this should be considered relative to adjacent agriculture. Dove populations depend upon agricultural crops for food. Such losses must be taken into account before justifying the area for wildlife conservation.

When the watertable is down 8 to 20 ft, the greatest water savings can be obtained by phreatophyte removal. These soils do not receive moisture except from surface water, and all measures of water conservation should be practiced.

When the water table is below 20 ft, invasion of saltcedar is much less prevalent and is usually killed by drought stress. These sites do not develop thickets, particularly preferred by wildlife populations. Cleared areas can readily be developed for picnicking and camping grounds.

Only rarely in the southwest are there areas which should be set aside by complete preservation. Most areas should be managed for multiple-use values that are consistent with the ecological potential of the site.

SUMMARY AND CONCLUSIONS

The Bureau of Reclamation, U.S. Army Corps of Engineers, U.S. Geological Survey, Bureau of Indian Affairs, U.S. Department of Agriculture, and various local and state governments have instituted programs for control and/or eradication of saltcedar and other phreatophytes.

The program of control has had limited success because the cost of control is relatively high for the economic benefits obtained. However, as the need for water increases with expanding populations, greater costs are justified and will likely be expended to control phreatophytes and riparian vegetation, and particularly, saltcedar.

REFERENCES

1. Horton, J. S. and Campbell, C. J., Management of Phreatophyte and Riparian Vegetation for Maximum Multiple-Use Values, Res. Paper RM-117, U.S. Forestry Service Department of Agriculture, Fort Collins, Colo., 1974.
2. Klingman, D. L., and Shaw, W. C., Using Phenoxy Herbicide Effectively, Farmers Bulletin No. 2183, U.S. Government Printing Office, Washington, D.C. 1962.
3. Pacific Southwest Interagency, Phreatophyte Subcommittee Meetings, Tucson, Ariz., 1970.
4. Robinson, T. W., Phreatophytes, Water-Supply Paper 1423, U.S. Geological Survey, Washington, D.C., 1958.
5. U.S. Department of the Interior, Pecos River Basin Water Salvage Project, New Mexico-Texas, Final Environmental Statement, FES 79-9, Southwest Region, Amarillo, Tex., 1979.

Appendixes

APPENDIX A

COMMON AND LATIN NAMES OF WEEDS

Agarita (*Mahonia trifoliolata* (Moric.) Fedde)
Alder (*Alnus* spp.)
Algae:
  Blue-green (many genera)
  Filamentous green (many genera) (*Pithophora* spp.)
Alligatorweed (*Alternanthera philoxeroides* (Mart.) Griseb.)
Arrowhead (*Sagittaria* spp.)
Ash (*Fraxinus* spp.)
Ash, white (*Fraxinus americana* L.)
Aster, woody (*Aster parryi* A. Gray)
Baccharis (*Baccharis* spp.) Barley, wild (*Hordeum leporinum* Link)
Barnyardgrass (*Echinochloa crusgalli* (L.) Beauv.)
Bayberry (*Myrica pensylvanica* Loisel.)
Beakrush (*Rhynchospora* spp.)
Bedstraw (*Galium* spp.)
Bedstraw, smooth (*Galium mollugo* L.)
Beech (*Fagus grandifolia* Ehrh.)
Bermudagrass (*Cynodon dactylon* (L.) Pers.)
Bindweed (*Convolvulus* spp.)
Bindweed, field (*Convolvulus arvensis* L.)
Birch (*Betula* spp.)
Bitterweed (*Helenium amarum* (Raf.) Rock)
Blackgum (*Nyssa sylvatica* Marsh.)
Bladderwort (*Ultricularia* spp.)
Bluegrass (*Poa* spp.)
Bluegrass, Kentucky (*Poa pratensis* L.)
Boxelder (*Acer negundo* L.)
Brackenfern (*Pteridium* spp.)
Bramble (*Rubus* sp.)
Brome, downy (*Bromus tectorum* L.)
Brome, Scotch (*Cytisus scoparius* (L.) Link)
Bromegrass (*Bromus* spp.)
Bromegrass, downy (*Bromus tectorum* L.)
Broomsedge (*Andropogon virginicus* L.)
Broomweed, common (*Guitierrezia dracunculoides* (DC.) Blake)
Buckbrush (*Symphoricarpos* spp.)
Buckwheat, wild (*Polygonum convolvulus* L.)
Bulrush (*Scirpus spp.)*
Burclover (Medicago polymorpha var. vulgaris (Benth.) Shinners)
Burdock (*Articum* spp.)
Burhead (*Echinodorus cordifolius* (L.) Griseb.)
Buttonbush (*Cephalanthus occidentalis* L.)
Cabomba (*Cabomba caroliniana* Gray)
Cactus, cholla (*Opuntia* spp.)
Canarygrass, reed (*Phalaris arundinacea* L.)
Cane, wild (*Sorghum bicolor* (L). Moench)
Carpetweed (*Mollugo verticillata.* L.)
Carrot, wild (*Daucus carota* L.)
Catclaw (*Schrankia nuttallii* (DC.) Standl.)
Cattail (*Typha* spp.)
Ceanothus (*Ceanothus* spp.)
Celery, wild (*Apium leptophyllum* (Pers.) F. Muell.)
Chara (*Chara* spp.)
Cheat (*Bromus secalinus* L.)
Cherry (*Prunus* spp.)
Chestnut (*Castanea* spp.)
Chickweed (*Stellaria media* (L.) Cyrillo)
Chicory (*Cichorium intybus* L.)
Chinaberry, wild (*Melia azedarach* L.)
Chinkapin (*Castanea* spp.)
Clover (*Trifolium* spp.)
Clover, white (*Trifolium repens* L.)
Cockle, corn (*Agrostemma githago* L.)
Cockle, cow (*Saponaria vaccaria* L.)
Cocklebur (*Xanthium* spp.)
Coffeebean (See *Sesbania*)
Coontail (*Ceratophyllum* spp.)
Cottonwood (*Populus* spp.)
Crabgrass (*Digitaria* spp.)
Crabgrass, hairy (*Digitaria villosa* (Walt.) Pers.)
Crabgrass, smooth (*Digitaria ischaemum* Schreb.- Muhl.)
Cress, hoary (*Cardaria draba* (L.) Desv.)

Crowfoot, water (*Ranunculus* spp.)
Cutgrass (*Leersia* spp.)
Daisy (*Chrysanthemum* spp.)
Dallisgrass (*Paspalum dilatatum* Poir.)
Dandelion (*Taraxacum officinale* Weber)
Deathcamas (*Zigadenus* spp.)
Dock, curly (*Rumex crispus* L.)
Dodder (*Cuscuta* spp.)
Dogbane (*Apocynum* spp.)
Dogfennel (*Eupatorium capillifolium* (Lam.) Small)
Dogwood (*Cormus* spp.)
Ducksalad (*Heteranthera limosa* (Sw.) Willd.)
Duckweed (*Lemna* spp.)
Elbowbush (*Forestiera pubescens* Nutt.)
Elderberry (*Sambucus* spp.)
Elm (*Ulmus* spp.)
Elodea (*Elodea* spp.)
Fescue, red (*Festuca rubra* L.)
Fescus, tall (*Festuca* spp.)
Fimbristylis (*Fimbristylis* spp.)
Flatsedge (*Cyperus* spp.)
Foxtail, giant (*Setaria faberii* Herrm.)
Foxtail, green (*Setaris viridis* (L.) Beauv.)
Foxtail, yellow (*Setaria glauca* (L.) Beauv.)
Garlic, wild (*Allium vineale* L.)
Goldenrod (*Solidago* spp.)
Goosegrass (*Eleusine indica* (L.) Gaertn.)
Gooseweed (*Sphenoclea zeylanica* Gaertn.)
Gorse (*Ulex europaeus* L.)
Groundcherry (*Physalis* spp.)
Hackberry (*Celtis* spp.)
Halogeton (*Halogeton glomeratus* (M. Bieb.) C. A. Mey)
Haw, red (*Crataegus* spp.)
Henbit (*Lamium amplexicaule* L.)
Hickory (*Carya* spp.)
Honeylocust (*Gleditsia triacanthos* L.)
Horehound (*Marrubium vulgare* L.)
Horsenettle (*Solanum carolinense* L.)
Horsetail rush (*Equisetum* spp.)
Huisache (*Acacia farnesiana* (L.) Willd.)
Hydrangea (*Hydrangea* spp.)
Indigo, curly (*Aeschynomene virginica* (L.) BSP.)
Ironweed (*Vernonia* spp.)
Ivy, ground (*Glechoma hederacea* L.)
Ivy, poison (*Rhus radicans* L.)
Johnsongrass (*Sorghum halepense* (L.) Pers.)
Jointvetch, northern (curly indigo) (*Aeschynomene indica* L.)
Knapweed, Russian (*Centaurea repens* L.)
Knawel (*Scleranthus annuus* L.)
Knotgrass (*Paspalum distichum* L.)
Knotweed (*Polygonum* spp.)
Lambsquarters (*Chenopodium* spp.)
Larkspur, low (*Delphinium nelsonii* Greene)
Larkspur, tall (*Delphinium barbeyi* Huth)
Lettuce, wild (*Lactuca* spp.)
Locoweed (*Astragalus* spp.)
Lotus, American (*Nelumbo lutea* (Willd.) Pers.)
Lupine, silvery (*Lupinus argenteus* Pursh)
Madrone (*Arbutus menziessi* Pursh)
Manzanita (*Arct staphylos* spp.)
Maple (*Acer* spp.)
Maple, red (*Acer rubrum* L.)
Maple, silver (*Acer saccharinum* L.)
Mescalbean (*Sophora secundiflora* (Ort.) Lag.)
Mesquite (*Prosopis* spp.)
Mesquite, honey (*Prosopis glandulosa* var. glandulosa (Torr.) Cockerell)
Mesquite, velvet (*Prosopis velutina* Woot.)
Mexicanweed (*Caperonia castaneaefolia* (L.) St. Hil.)
Milkvetch, timberline (*Astragalus* sp.)
Milkvetch, twogrooved (*Astragalus bisulcatus* (Hook.) Gray)
Morningglory (*Ipomoea* spp.)
Mulberry (*Morus* spp.)
Mullein (*Verbascum* spp.)
Mustard (*Brassica* spp.)
Mustard, tumble (*Sisymbrium altissimum* L.)
Naiad (*Najas* spp.)
Nightshade (*Solanum* spp.)
Nimblewill (*Muhlenbergia schreberi* J. F. Gmel.)
Nutsedge (*Cyperus* spp.)
Nutsedge, purple (*Cyperus rotundus* L.)
Nutsedge, yellow (*Cyperus esculentus* L.)
Oak (*Quercus* spp.)
Oak, blackjack (*Quercus marilandica* Muenchh.)
Oak, bur (*Quercus macrocarpa* Michx.)

Oak, pin (*Quercus palustris* Muenchh.)
Oak, poison (*Rhus toxicodendron* L.)
Oak, post (*Quercus stellata* Wangenb.)
Oak, red (*Quercus rubra* L.)
Oak, shinnery (*Quercus havardii* Rydb.)
Oak, water (*Quercus nigra* L.)
Oak, white (*Quercus alba* L.)
Oak, willow (*Quercus phellos* L.)
Oats, wild (*Avena fatua* L.)
Onion, wild (*Allium canadense* L.)
Orchardgrass (*Dactylis glomerata* L.)
Osageorange (*Maclura pomifera* (Raf.) Schneid.)
Palmetto (*Serenoa repens* (Bartr.) Small)
Paragrass (*Panicum purpurascens* Raddi)
Pecan (*Carya* spp.)
Pennycress (*Thalspi* spp.)
Pennywort (*Hydrocotyle spp. umbellata* L.)
Pepperweed (*Lepidium* spp.)
Persimmon (*Diospyros* spp.)
Phragmites (*Phragmites communis* Trin.)
Pickerelweed (*Pontederia cordata* L.)
Pigweed (*Amaranthus* spp.)
Plantain, buckhorn (*Plantago lanceolata* L.)
Pondweed (*Potamogeton* spp.)
Pondweed, curlyleaf (*Potamogeton crispus* L.)
Poplar (*Populus* spp.)
Pricklyash (*Zanthoxylum americanum* Mill.)
Pricklypear (*Opuntia* spp.)
Princesplume (*Polygonum orientale* L.)
Puncturevine (*Tribulus terrestris* L.)
Purslane (*Portulaca oleracea* L.)
Quackgrass (*Agropyron repens* (L.) Beauv.)
Rabbitbrush (*Chrysothamnus* spp.)
Ragweed, bur (*Franseria discolor* Nutt.)
Ragweed, common (*Ambrosia artemisiifolia* L.)
Ragwort, tansy (*Senecio jacobaea* L.)
Redbud (*Cercis* spp.)
Redstem (*Anmania auriculata* Willd.)
Reed, common (*Phragmites communis* Trin.)
Reeds (*Phragmites* spp.)
Retama (*Parkinsonia aculeata* L.)
Rice, red (*Oryza sativa* L.)
Rocket, yellow (*Barbarea vulgaris* R. Br.)
Rushes (*Juncus* spp.)
Sagebrush, big (*Artemisia tridentata* Nutt.)
Sagebrush, low (*Artemisia arbuscula* Nutt.)
Sagebrush, sand (*Artemisia filifolia* Torr.)
Sagittaria, dwarf *Sagittaris sublata* (L.) Buchenau)
Saltcedar (*Tamarix pentandra* Pall.)
Sandbur (*Cenchrus* spp.)
Sassafras (*Sassafras* spp.)
Sedge, Nebraska (*Carex nebraskensis* Dewey)
Sedge, ripgut (*Carex lacustris* Willd.)
Sedges (*Carex* spp.)
Sesbania, hemp (*Sesbania exaltata* (Raf.) Cory)
Shepherdspurse (*Capsella bursa-pastoris* (L.) Medic.)
Signalgrass, broadleaf (*Brachiaria platyphylla* (Griseb.) Nash)
Smartweed (*Polygonum* spp.)
Sneezeweed (*Helenium* spp.)
Sneezeweed, orange (western (*Helenium hoopesii* Gray)
Sorrel, red (*Rumex acetosella* L.)
Sowthistle (*Sonchus* spp.)
Spatterdock (*Nuphar advena* (Ait.) Ait. f.)
Spicebush (*Lindera benzoin* (L.) Blume)
Spikerush (*Eleocharis* spp.)
Spirea, narrowleaf (*Spiraea alba* DuRoi)
Sprangletop (*Leptochloa* spp.)
Spurge, leafy (*Euphorbia esula* L.)
St. Johnswort (*Hypericum perforatum* L.)
Strawberry, wild (*Fragaria* spp.)
Sumac (*Rhus* spp.)
Sweetgum (*Liquidambar styraciflura* L.)
Sycamore (*Platanus* spp.)
Tansymustard (*Descurainia pinnata* (Walt.) Britt.)
Tarweed (*Madia* spp.) Tasajillo (*Opuntia leptocaulis* DC.)
Thistle, Canada (*Cirsium arvense* (L.) Scop.)
Thistle, Russian (*Salsola kali* L. var. tenuifolia Tausch)
Timothy (*Phleum pratense* L.)
Toadflax, Dalmatian (*Linaria dalmatica* Mill.)

Tree-of-heaven (*Ailanthus altissima* (Mill.) Swingle)
Trumpetvine (*Campsis radicans* (L.) Seem.)
Tules (*Scirpus* spp.) (See Bulrush)
Velvetleaf (*Abutilon theophrasti* Medic.)
Violet (*Viola* spp.)
Walnut (*Juglans* spp.)
Waterchestnut (*Trapa natans* L.)
Watercress (*Nasturtium officinale* R. Br.)
Waterfern (*Salvinia* spp.)
Waterhemlock (*Cicuta* spp.)
Waterhyacinth (*Eichhornia crassipes* (Mart.) Solms.)
Waterhyssop (*Bacopa caroliniana* (Walt.) Robins.)
Waterlettuce (*Pistia stratiotes* L.)
Waterlily (*Nymphaea* spp.)
Waterlily, white (*Nymphaea tuberosa* Paine)
Watermilfoil (*Myriophyllum* spp.)
Waterplantain (*Alisma* spp.)
Waterprimrose (*Jussiaea* spp.)
Watershield (*Brasenia schreberi* Gmel.)
Waterstargrass (*Heteranthera dubia* (Jacq.) MacM.)
Wildrye (*Elymus* spp.)
Willow (*Salix* spp.)
Woodsorrel (*Oxalis* spp.)
Yerba-de-tago (*Eclipta alba* (L.) Hassk.)

APPENDIX B
COMMON AND CHEMICAL NAMES OF HERBICIDES

Common Name or Designation	Chemical Name
acrolein	acrolein
alachlor	2-chloro-2′,6′-diethyl-*N*-(methoxymethyl)acetanilide
ametryn	2-(ethylamino)-4-(isopropylamino)-6-(methylthio)-*s*-triazine
amitrole	3-amino-*s*-triazole
AMS	ammonium sulfamate
asulam	methyl sulfanilylcarbamate
atrazine	2-chloro-4-(ethylamino)-6-(isopropylamino)-*s*-triazine
barban	4-chloro-2-butynyl *m*-chlorocarbanilate
benefin	*N*-butyl-*N*-ethyl-α,α,α-trifluoro-2,6-dinitro-*p*-toluidine
bensulide	*O,O*-diisopropyl phosphorodithioate *S*-ester with *N*-(2-mercaptoethyl)benzenesulfonamide
bentazon	3-isopropyl-1 *H*-2,1,3-benzothiadiazin-(4) 3*H*-one 2,2-dioxide
benzadox	(benzamidooxy)acetic acid
benzipram	3,5-dimethyl-*N*-(1-methylethyl)-*N*-(phenylmethyl)benzamide
bifenox	methyl 5-(2,4-dichlorophenoxy)-2-nitrobenzoate
bromacil	5-bromo-3-*sec*-butyl-6-methyluracil
bromoxynil	3,5-dibromo-4-hydroxybenzonitrile
butachlor	*N*-(butoxymethyl)-2-chloro-2′,6′-diethylacetanilide
butam	2,2-dimethyl-*N*-(1-methylethyl)-*N*-(phenylmethyl)propanamide
buthidazole	3-[5-(1,1-dimethylethyl)-1,3,4-thiadiazol-2-yl]-4-hydroxy-1-methyl-2-imidazolidinone
butralin	4-(1,1-dimethylethyl)-*N*-(1-methylpropyl)-2,6-dinitrobenzenamine
butylate	*S*-ethyldiisobutylthiocarbamate
cacodylic acid	hydroxydimethylarsine oxide
carbetamide	*D*-*N*-ethyllactamide carbanilate (ester)
CDAA	*N-N*-diallyl-2-chloroacetamide
CDEC	2-chloroallyl diethyldithiocarbamate
chloramben	3-amino-2,5-dichlorobenzoic acid
chlorbromuron	3-(4-bromo-3-chlorophenyl)-1-methoxy-1-methylurea
chloroxuron	3-[*p*-(*p*-chlorophenoxy)phenyl]-1,1-dimethylurea
chlorpropham	isopropyl *m*-chlorocarbanilate
cisanilide	*cis*-2,5-dimethyl-*N*-phenyl-1-pyrrolidinecarboxamide
CMA	calcium methanearsonate
cyanazine	2-[[4-chloro-6-(ethylamino)-*s*-triazin-2-yl]amino]-2-methylpropionitrile
cycloate	*S*-ethyl *N*-ethylthiocyclohexanecarbamate
cycluron	3-cyclooctyl-1,1-dimethylurea
cyperquat	1-methyl-4-phenylpyridinium
cyprazine	2-chloro-4-(cyclopropylamino)-6-(isopropylamino)-*s*-triazine
cyprazole	*N*-[5-(2-chloro-1,1-dimethylethyl)-1,3,4-thiadiazol-2-yl] cyclopropanecarboxamide
cypromid	3′,4′-dichlorocyclopropanecarboxanilide
dalapon	2,2-dichloropropionic acid
dazomet	tetrahydro-3,5-dimethyl-2*H*-1,3,5-thiadiazine-2-thione
DCPA	dimethyl tetrachloroterephthalate
desmedipham	ethyl *m*-hydroxycarbanilate carbanilate (ester)
desmetryn	2-(isopropylamino)-4-(methylamino)-6-(methylthio)-*s*-triazine
diallate	*S*-(2,3-dichloroallyl)diisopropylthiocarbamate
dicamba	3,6-dichloro-*o*-anisic acid
dichlobenil	2,6-dichlorobenzonitrile
dichlorprop	2-(2,4-dichlorophenoxy) propionic acid
difenzoquat	1,2-dimethyl-3,5-diphenyl-1*H*-pyrazolium
dinitramine	*N*′,*N*′-diethyl-α,α,α-trifluoro-3,5-dinitrotoluene-2,4-diamine
dinoseb	2-*sec*-butyl-4,6-dinitrophenol
diphenamid	*N,N*-dimethyl-2,2-dipheylacetamide
dipropetryn	2-ethylthio-4-6-bis-isopropylamino-*s*-triazine
diquat	6,7-dihydrodipyrido[1,2-α:2′,1′-*c*] pyrazinediium ion
diuron	3-(3,4-dichlorophenyl)-1,1-dimethylurea
DNOC	4,6-dinitro-*o*-cresol
DSMA	disodium methanearsonate
endothall	7-oxabicyclo[2.2.1]heptane-2,3-dicarboxylic acid
EPTC	*S*-ethyl dipropylthiocarbamate
erbon	2-(2,4,5-trichlorophenoxy)ethyl 2,2-dichloropropionate
ethalfluralin	*N*-ethyl-*N*-(2-methyl-2-propenyl)-2,6-dinitro-4-(trifluoromethyl)benzenamine
ethiolate	*S*-ethyl diethylthiocarbamate
fenac	(2,3,6-trichlorophenyl)acetic acid
fenuron	1,1-dimethyl-3-phenylurea
fenuronTCA	1,1-dimethyl-3-phenylurea mono(trichloroactetate)
fluchloralin	*N*-(2-chloroethyl)-2,6-dinitro-*N*-propyl-4-(trifluoromethyl)aniline
fluometuron	1,1-dimethyl-3-(α,α,α-trifluoro-*m*-totyl)urea
fluorodifen	*p*-nitrophenyl α,α,α-trifluoro-2-nitro-*p*-tolyl ether
glyphosate	*N*-(phosphonomethyl)glycine
hexaflurate	potassium hexafluoroarsenate
ioxynil	4-hydroxy-3,5-diiodobenzonitrile
isopropalin	2,6-dinitro-*N,N*-dipropylcumidine
karbutilate	*tert*-butylcarbamic acid ester with 3(*m*-hydroxyphenyl)-1, 1-dimethylurea
lenacil	3-cyclohexyl-6.7-dihydro-1*H*-cyclopentapyrimidine-2,4(3*H*,5*H*)-dione
linuron	3-(3,4-dichlorophenyl)-1-methoxy-1-methylurea
MAA	methanearsonic acid
MAMA	monoammonium methanearsonate
MCPA	[(4-chloro-*o*-tolyl)oxy]acetic acid
MCPB	4-[(4-chloro-*o*-tolyl)oxy]butyric acid
mecoprop	2-[(4-chloro-*o*-tolyl)oxy]propionic acid

mefluidide	*N*-[2,4-dimethyl-5-[[(trifluoromethyl)sulfonyl]amino]phenyl]acetamide
metham	sodium methyldithiocarbamate
methazole	2-(3,4-dichlorophenyl)-4-methyl-1,2,4-oxadiazolidine-3,5-dione
metolachlor	2-chloro-*N*-(2-ethyl-6-methylphenyl)-*N*-(2-methoxy-1-methylethyl)acetamide
metribuzin	4-amino-6-*tert*-butyl-3-(methylthio)-*as*-triazin-5(4*H*)one
MH	1,2-dihydro-3,6-pyridazinedione
molinate	S-ethyl hexahydro-1*H*-azepine-1-carbothioate
monolinuron	3-(*p*-chlorophenyl)-1-methoxy-1-methylurea
monuron	3-(*p*-chlorophenyl)-1,1-dimethylurea
monuronTCA	3-(*p*-chlorophenyl)-1,1-dimethylurea mono(trichloroacetate)
MSMA	monosodium methanearsonate
napropamide	2-(α-naphthoxy)-*N,N*-diethylpropionamide
naptalam	*N*-1-naphthylphthalamic acid
neburon	1-butyl-3-(3,4-dichlorophenyl)-1-methylurea
nitralin	4-(methylsulfonyl)-2,6-dinitrol-*N,N*-dipropylaniline
nitrofen	2,4-dichlorophenyl-*p*-nitrophenyl ether
nitrofluorfen	2-chloro-1-(4-nitrophenoxy)-4-(trifluoromethyl)benzene
norea	3-(hexahydro-4,7-methanoindan-5-yl)-1,1-dimethylurea
norflurazon	4-chloro-5-(methylamino)-2-(α,α,α-trifluoro-*m*-tolyl)-3(2*H*)-pyridazinone
oryzalin	3,5-dinitro-N^4,N^4-dipropylsulfanilamide
oxadiazon	2-*tert*-butyl-4-(dichloro-5-isopropoxyphenyl)-Δ^2-1,3,4-oxadiazolin-5-one
oxyfluorfen	2-chloro-1-(3-ethoxy-4-nitrophenoxy)-4-(trifluoromethyl)benzene
paraquat	1,1′-dimethyl-4,4′-bipyridinium ion
PBA	chlorinated benzoic acid
pebulate	S-propyl butylethylthiocarbamate
pendimethalin	*N*-(1-ethylpropyl)-3,4-dimethyl-2,6-dinitrobenzenamine
perfluidone	1,1,1-trifluoro-*N*-[2-methyl-4-(phenylsulfonyl)phenyl] methanesulfonamide
phenmedipham	methyl *m*-hydroxycarbanilate *m*-methylcarbanilate
picloram	4-amino-3,5,6-trichloropicolinic acid
procyazine	2-[[4-chloro-6-(cyclopropylamino)-1,3,5-triazine-2-yl]amino]-2-methylpropanenitrile
profluralin	*N*-(cyclopropylmethyl)-α,α,α-trifluoro-2,6-dinitro-*N*-propyl-*p*-toluidine
prometone	2,4-bis(isopropylamino)-6-methoxy-*s*-triazine
prometryn	2,4-bis(isopropylamino)-6-(methylthio)-*s*-triazine
pronamide	3,5-dichloro(*N*-1,1-dimethyl-2-propynyl)benzamide
propachlor	2-chloro-*N*-isopropylacetanilide
propanil	3′4′-dichloropropionanilide
propazine	2-chloro-4,6-bis(isopropylamino)-*s*-triazine
propham	isopropyl carbanilate
prosulfalin	N-[[4-dipropylamino)-3,5-dinitrophenyl]sulfonyl]-S,S-dimethylsulfilimine
prynachlor	2-chloro-*N*-(1-methyl-2-propynyl)acetanilide
pyrazon	5-amino-4-chloro-2-phenyl-3(2*H*)-pyridazinone
secbumeton	*N*-ethyl-6-methoxy-*N*′(1-methylpropyl)-1,3,5-triazine-2,4-diamine
siduron	1-(2-methylcyclohexyl)-3-phenylurea
silvex	2-(2,4,5-trichlorophenoxy)propionic acid
simazine	2-chloro-4,6-bis(ethylamino)-*s*-triazine
simetryn	2,4-bis(ethylamino)-6-(methylthio)-*s*-triazine
TCA	trichloroacetic acid
tebuthiuron	*N*-[5-(1,1-dimethylethyl)-1,3,4-thiadiazol-2-yl]-*N,N*′-dimethylurea
terbacil	3-*tert*-butyl-5-chloro-6-methyluracil
terbuchlor	*N*-(butoxymethyl)-2-chloro-*N*-[2-(1,1-dimethylethyl)-6-methylphenyl]acetamide
terbuthylazine	2-(*tert*-butylamino)-4-chloro-6-(ethylamino)-*s*-triazine
terbutol	2,6-di-*tert*-butyl-*p*-tolyl methylcarbamate
terbutryn	2-(*tert*-butylamino)-4-(ethylamino)-6-(methylthio)-*s*-triazine
triallate	S-(2,3,3-trichloroallyl)diisopropylthiocarbamate
triclopyr	[(3,5,6-trichloro-2-pyridinyl)oxy]acetic acid
trifluralin	α,α,α-trifluoro-2,6-dinitro-*N,N*-dipropyl-*p*-toluidine
trimeturon	1-(*p*-chlorophenyl)-2,3,3-trimethylpseudourea
2,3,6-TBA[b]	2,3,6-trichlorobenzoic acid
2,4-D	(2,4-dichlorophenoxy)acetic acid
2,4-DB	4-(2,4-dichlorophenoxy)butyric acid
2,4-DEP	tris[2-(2,4-dichlorophenoxy)ethyl] phosphite
2,4,5-T	(2,4,5-trichlorophenoxy)acetic acid
vernolate	S-propyl dipropylthiocarbamate

[a] Herbicides no longer in use in USA are omitted. Complete listing, including these, is in WEEDS 14(4), 1966 and in Weed Science 23(1), 1975.

[b] This herbicide usually is available as mixed isomers. When possible, the isomers should be identified, the amount of each isomer in the mixture specified and the source of the experimental chemicals given.

GLOSSARY OF BIOAGRONOMIC TERMS

The definitions and explanations in this glossary apply to words as they are used in context. Most words have more comprehensive meanings.

Absorb — To take a pesticide into a plant, animal, or the soil.
Absorptive clay — A special type of clay powder which can take up chemicals and hold them. It is sometimes used to clean up spills of pesticides.
Accumulative pesticides — Those chemicals which tend to build up in animals or the environment.
Acid — A very sharp, sour liquid which is frequently toxic in concentrated form.
Acid equivalent (ae) — The theoretical yield of parent acid from an active ingredient. It is used instead of or in addition to the active ingredient for certain herbicides.
Acre — 43,560 ft^2. An area of land about 209 ft long by 209 ft wide.
Activated charcoal — Very finely ground, high-quality charcoal which absorbs liquids and gasses easily.
Active ingredient (ai) — The part of the pesticide which will kill pests or prevent damage by them. Usually it is the same as the "technical" material in the formulated product.
Actual dosage — The amount of active ingredient (not formulated product) which is applied to an area or other target.
Acute poisoning — Severe poisoning which occurs after one exposure to a pesticide.
Acute toxicity — How poisonous a pesticide is to an animal (or man) after a single dose or exposure.
Adhesive — An adjuvant which helps the pesticide stick to the treated surface.
Adjuvant-additive — Something added to the pesticide mixture to help the active ingredient do a better job. Examples: wetting agent, spreader, adhesive, emulsifying agent, penetrant.
Adsorbed — Held so tightly that the herbicide is rendered inactive or only slowly effective. The principle adsorbing agents of the soil are its inorganic (clay) and organic (humus) colloids.
Aeolian loess — Soils formed by accumulation of wind-blown particles.
Aerosol — Pesticide chemical stored in a container under pressure. The pesticide is driven through a fine opening by an inactive gas under pressure when the nozzle is triggered.
Agitate — To keep a pesticide chemical mixed up; to keep it from settling or separating in the spray tank.
Algal blooms — A readily visible concentrated growth or aggregation of algae in water; commonly accepted to be a concentration of cells greater than 500 per milliliter.
Aliphatic materials — Chemically, those that have an open-chain molecular structure. As herbicides, they are less toxic to plants than aromatic compounds.
Alkali — Opposite of an acid; it is usually toxic in concentrated form.
Angiosperms — Taxonomic unit of plants which includes all of the potentially flowering plants in which the ovules and seeds, when formed, form inside a closed carpel system collectively known as fruits when mature.
Anionic surfactant — One that has a negative charge and performs best in cold water and soft water. Most wetting agents and detergents and some emulsifiers are anionic.
Annual — A plant that completes its life cycle from seed in 1 year.
Antidote — A treatment given by a doctor to reduce the effects of pesticide poisoning.

Antisiphoning device — A small piece of equipment attached to the filling hose to prevent fill water from draining back into the water source. Example: check valve.

Application — Putting a pesticide on or in plants, animals, buildings, soil, air, water, or other target to kill pests or prevent damage by them.

Aquatic macrophytes — Aquatic plants which may be seen with the naked eye, e.g., aquatic mosses, ferns, and rooted plants.

Aquatic vascular plants — Aquatic plants which have specialized conductive tissues.

Aquatic weeds — Weeds that grow in water, either on top or under the surface.

Aromatic oils and solvents — Chemically, those that have unsaturated molecular structure. As herbicides, they are generally more toxic to plants than aliphatic materials.

Artificial respiration — First aid for someone who has stopped breathing, by blowing air into his lungs or applying pressure to his back to start breathing again.

Attractive nuisance — A legal term for any object which might attract children or other persons to it and then might injure or hurt them as a result. Examples: sprayers and empty pesticide containers.

Biennial — A plant that completes its life cycle in 2 years. The first year it produces leaves and stores food. The second year it produces fruits and seeds.

Biomass — The total amount (weight) of living material in a given habitat.

Botanical plant name — A scientific name made up of the genus and species. Sometimes the variety or subspecies is included. The botanical name is more reliable and more universal than the common names.

Biological control — Pest control without the use of chemicals. Parasites, predators, diseases, etc. are used to control pests.

Boom — Several nozzles joined together by sections of pipe or tubing to apply pesticides over a wider area at one time.

Broadcast application — Uniform distribution of an herbicide over an entire area.

Broadcast plants — Botanically, those classified as dicotyledons; morphologically, those that have broad, usually compound leaves.

Broad spectrum pesticides — A general-purpose pesticide with a wide range of uses. They are effective when several different pests are a problem to control.

Calculate — Do some arithmetic; work with numbers, determine, figure out, i.e., rate per acre.

Calibrate (calibration) — To figure or measure how much pesticide will be applied by the equipment to the target area.

Canceled — A pesticide use that is no longer registered as a legal use by the Environmental Protection Agency (EPA). Remaining stocks can be used by order of the Administrator, EPA.

Carrier — The liquid or solid material added to a chemical compound to facilitate its application in the field.

Cartridge — The cylinder-shaped part of the respirator which absorbs the fumes and vapors from the air before you breathe them. It should be replaced often.

Cation exchange — The exchange of positive ions (H, Ca, Mg, Na, and NH_4) from clay particles for other cations. Soils are able to filter out salts in much the same way a water softener removes them. Some soils have a larger capacity than others for doing this. Such soils can adsorb or filter out and hold large amounts of an herbicide so that it is not immediately effective.

Caution — A signal word used on labels of pesticides to alert users that the pesticide is slightly toxic.

Chemical — Used here to mean a "pesticide" chemical.

Chemical name — Scientific name telling the contents or formula of the active ingredients of the pesticide.

Chemically inactive — Will not easily react with any other chemical or object. Examples: talc and clay.

Chemical reaction — When two or more substances are combined and as a result undergo a complete change to make new substances or materials.

Chronic poisoning — Poisoning which occurs as a result of small, repeated doses of pesticide over a long period of time.

Chronic toxicity — How poisonous a pesticide is to an animal (or man) after small, repeated doses over a period of time.

Circulate — To move completely through something in a path that returns to the starting point.

Commercial — A job or business whose purpose is to make money or earn a profit.

Common plant name — An English name in common use. A plant may be known by several different common names, and one common name may be used for different plants if different parts of the country. Many common names are local.

Compatible pesticides — Compounds or formulations that can be mixed and applied together without undesirably altering their separate effects.

Concentration — The amount of active ingredient or acid equivalent in a given volume of liquid or in a given weight of dry material.

Condemnation — The act of removing a crop or product which does not meet legal standards for tolerances on foods and thus is not to be sold.

Contact herbicide — One that kills primarily by contact with plant tissue rather than as a result of translocation. Toxic upon contract with target or nontarget species.

Contaminate — The addition of an unwanted material (often a pesticide) where it could do harm or damage or interfere with natural processes.

Contract application — Someone who contracts to apply pesticides.

Control program — An organized program to keep down the number of pests in an area.

Corrosion — The effect of being worn down or eaten away slowly.

Corrosive herbicide — A type of herbicide containing a strong acid or base which will severely burn the skin, mouth, stomach, etc. or corrode application equipment.

Cross contamination — When one pesticide gets into or mixes with another pesticide accidentally — usually occurs in pesticide container or in a poorly cleaned sprayer.

Days to harvest — The least number of days between the last pesticide application and the harvest date, as set by law.

Decontaminate — To remove or break down the unwanted material (usually pesticide) so it cannot do any harm or damage.

Defoliant — A type of pesticide which causes the leaves of a plant to drop off.

Degrade — Break down, decompose by biologic or other means.

Degree of exposure — The amount or extent to which a person has been in contact with a toxic pesticide.

Deposit — The pesticide on the leaves or skin or other surface immediately after pesticide application.

Dermal toxicity — How poisonous a pesticide is to an animal when absorbed through the skin.

Desiccant — A type of pesticide which draws moistures (liquids) from a plant or plant part causing it to wither and die.

Detergent — A chemical (not soap) having the ability to remove soil or grime. Household detergents can be use as surfactants in herbicide sprays.

Detritus — Fragments of dead organic matter in water.

Diluent — Any liquid or solid material that dilutes an active ingredient in the preparation of a formulation.

Disposal — The act or process of discarding or throwing away a pesticide.

Diversity — Pertaining to the variety of species within a given association of organisms. Areas of high diversity are characterized by a great variety of species; areas of low diversity contain few species, although large numbers of individuals may be present.

Dormant spray — Pesticide application made before trees and other plant life begin to leaf out in the spring.

Dose (dosage) — The portion or amount of pesticide mixture which is applied to the target.

Downwind — On the side from which the prevailing wind is blowing towards.

Drawdown reservoir — A reservoir which experiences marked water level fluctuations over a time period, naturally, or by specific design.

Drift — The movement by wind and air currents of droplets or particles of a pesticide from the target area to an area not intended to be treated.

Dust — A finely ground, dry mixture containing a small amount of pesticide and an inert carrier such as talc or clay. The dust particles are of many different sizes.

Ecology — Study of the relationship between a plant or animal and its surroundings.

Emergent plants — Aquatic macrophytes that are rooted in the sediments of a water body, but with vegetative parts which project above the water surface.

Emulsifiable concentrates — Usually liquids in which the chemical is dissolved in one or more water-insoluble solvents such as oil or benzene to which an emulsifier is added.

Emulsifier — A surface active material that facilitates the suspension of one liquid in another.

Emulsion — The suspension of one liquid as minute globules in another liquid; e.g., oil dispersed in water.

Encapsulation — Method of disposal of pesticides and pesticide containers by sealing them in a sturdy, waterproof container so the contents cannot possibly get out. Also a method of formulating pesticides.

Environment — Surroundings — usually water, air, soil, plants, and wildlife.

EPA registration number — A number assigned by EPA to a product when it is registered that must appear on all labels for that product. It will appear as "EPA Reg. No." or "EPA Registration No." followed by the company number and product number.

Eutrophication — The process of enrichment with nutrients, leading to increased organic production and possibly algal blooms.

Evaporate — To form a gas and disappear into the air.

Exemption — Exception to the registered use of a pesticide.

Exposure — The amount of contact with a pesticide.

Extruded — A process in which a powdered carrier mixed with the herbicide is moistened until it becomes plastic and then is forced or expelled as rods. These rods are dried, ground, and screened to the required screen mesh for a granular formulation. This process increases internal pore volume and surface area.

Face shield — A transparent piece of protective equipment used by a pesticide applicator to protect his face from exposure to pesticides.

Family — Collection of species and genera of assumed close relationship; usually plant families end with the letters -aceae or -ae; see Taxon.

Finite tolerance — The maximum amount of pesticide which can remain on food or feed crops at harvest after the pesticide has been directly applied to the crops.

Flowable — Very finely ground solid materials of pesticide which are mixed in a liquid carrier.

Fluid — Liquid.

Foaming agent — Chemical substance which causes the pesticide mixture to form a thick foam. It is used to reduce drift.

Fogger — Aerosol generator; a piece of pesticide equipment that breaks some pesticides into very fine droplets (aerosols or smokes) and blows or drifts the "fog" onto the target area.

Foliage — Leaves, needles, or blades of a plant.

Foliar sprays — Pesticides which are applied on the stems, leaves, needles, or blades of a plant.

Food chain — A way of describing how all animals depend on each other for food. It is a link between planteaters, plant and meat-eaters, and meat-eaters.

Formula — A brief way of writing a complicated idea by using abbreviations and symbols.

Formulation — A mixture of one or more pesticides plus other materials, such as carriers, diluents, etc., needed to make it safe and easy to store, dilute, and apply. The formulation is the form the pesticide is brought in; does not include tank mixes, adjuvants, etc.

Granular products — Formulations in which the chemical is impregnated on or in vermiculite, attaclay, or other suitable carriers and then formed into granules or pellets.

Grass — Botanically, any plant of the Gramineae family. Grasses are characterized by narrow leaves with parallel veins; by leaves composed of blade, sheath, and ligule; by jointed stems and fibrous roots; and by inconspicuous flowers usually arranged in spikelets.

Growth regulator — An organic substance effective in minute amounts for controlling or modifying plant processes.

Hard water — Water that contains certain minerals, usually calcium and magnesium sulfates, chlorides, or carbonates, in solution in amounts that cause a curd or precipitate instead of a lather when soap is added. Generally defined as containing 322 ppm in terms of calcium carbonate. Very hard water may cause precipitates in some herbicidal sprays.

Hazard — The risk of danger; the chance that danger or harm will come to the applicator, bystanders, consumers, livestock, wildlife or crops, etc.

Herbaceous plant — A vascular plant that does not develop woody tissue. It dies down each year.

Herbarium (plural: herbaria) — A permanent collection of identified, dried, preserved, and mounted plants for research and plant identification reference purposes; ideally, specimens each have identification data including collector, date of collection, name and taxonomic authority, site of collection, together with names of associated species and ecological data.

Herbicide — A chemical used for killing or interrupting the normal growth of plants.

Hydraulic agitator — A device which keeps the tank mix from settling out by means of water flow under pressure.

Hydraulic sprayer — A machine which applies pesticides by using water at high pressure and volume to deliver the pesticide to the target.

Illegal residue — A quantity of pesticide remaining on the crop at harvest which is either above the set tolerance or which is not allowed on the crop at all.

Inactive — Will not react chemically with anything; not involved in the pesticide action.

Incinerator — A high heat furnace or burner which reduces everything to ashes and vapors or nonharmful residues.

Incompatible — Not able to be mixed or used together.

Incorporate — To work or blend a pesticide into the soil completely.

Inert ingredients — Inactive part of a pesticide/formulation; any material in a pesticide mixture which would not prevent damage or destroy pests if used by itself.

Infestation — Any pests found in an area or place where they are not desirable.

Ingredient statement — The part of the label on a pesticide container which gives the name and amount of each pesticide chemical and the amount of inactive material in the mixture.

Inhalation — To take air into the lungs, to breath in.

Inhalation toxicity — How poisonous a pesticide is to man or an animal when breathed in through the lungs.

Inject — To force a pesticide chemical into a plant, animal, building, or the soil.

Integrated control — A system in which two or more methods are used to control a pest. These methods may include cultural practices, natural enemies, and selective pesticides.

Interval — The time period between two pesticide applications or between the last pesticide application and harvest.

Invert emulsion — One in which oil is the continuous phase and water is dispersed in it.

Ionic surfactant — One that ionizes or dissociates in water.

Irritating — Annoying. Making an animal (or person) uncomfortable by burning, stinging, tickling, making the eyes water, etc.

Isomers — Two or more substance having the same chemical composition, but different properties.

Jointly liable — Two or more persons or companies would share legal responsibility for negligence.

Kg or kilogram — A unit of weight in the metric system equal to 2.2 lb.

Label — The printed material attached to or part of a pesticide container.

LC_{50} — The concentration of a pesticide in air which would kill half of the test animals exposed to it. The lower the LC_{50} value, the more poisonous the pesticide. It is often used as the measure of acute inhalation toxicity.

LD_{50} — The dose or amount of a pesticide which would kill half or a large number of test animals if eaten or absorbed through the skin. The lower the LD_{50} value, the more poisonous the pesticide. LD_{50} values are the commonly used measure of acute oral and acute dermal toxicity.

Lethal — Deadly toxic, i.e., causing death of target or nontarget species.

Littoral — The shallow zone of a water body where light penetrates to the bottom; the zone of rooted aquatic vegetation.

Low-volatile ester — Chemically, an ester with a heavy molecular weight such as the butoxy-ethanol, iso-octyl, or propylene glycol butyl ether esters. Low-volatile esters do not include the methyl, ethyl, propyl, isopropyl, butyl, amyl, and pentyl esters. Biologically, an ester that is less liable than the high-volatile esters to injure plants by vapor activity.

Material — A substance, often used to mean a pesticide chemical.

Maximum dosage — The largest amount of a pesticide chemical that is safe to use without resulting in excess residues or damage to whatever is being protected.

Mechanical agitator — A device which keeps the pesticide and any additives thoroughly mixed in the spray tank by padding, swirling, or stirring.

Metric — A system of measurement which is used by most of the world except the U.S. and Canada and which is used in scientific work. It uses meters, grams, and liters as units.

Negligible residue — A tolerance which is set on a food or feed crop which will have a very small amount of pesticide at harvest as a result of indirect contact with the chemical.

Nonaccumulative — Will not build up in an animal's body or in the environment.

Nonionic surfactant — Chemically inert.

Nonlabeled — Use or method which is not written on the pesticide label and therefore is not legal.

Nonpersistent — Only lasts a short time (a few weeks or less) after being applied, breaks down rapidly in the environment.

Nonselective pesticide — A pesticide chemical that will control a wide range of pests.

Nontarget — Any plant, animal, or other organism that a pesticide application is not aimed at, but may accidently be injured by the chemical.

Nonvolatile — A pesticide chemical that does not evaporate (turn into a gas or vapor) at normal temperatures.

Operating speed — The steady rate which your pesticide sprayer is moving along the ground — usually measured in miles per hour or feet per minute.

Oligotrophic — Water low in plant nutrients.

Organic matter — Plant or animal remains which are partially decomposed. See detritus.

Parts per million or ppm — The amount of pesticide that remains on or in a plant, animal, food, or feed crop after treatment (residue) is often measured in parts per million.

Penetrant — A kind of additive or adjuvant which aids the pesticide to get through the outer surface (leaf, root, skin) and into the plant.

Percent by weight — The amount of acutal pesticide chemical in a mixture based on its weight compared to the weight of the whole mixture.

Example: 1 lb of actual pesticide plus 3 lb of other materials would give you a 25% pesticide by weight in the mixture.

Perennial — A plant that normally lives for more than 2 years. Trees and shrubs are perennials.

Persist — To stay for a period of time; to remain.

Pest — An unwanted organism (animal, plant, bacteria, fungus, virus, etc.)

Pesticide — A chemical or other agent that will destroy a pest or protect something from a pest.

Pesticide chemical — Term used to describe a pesticide which is a chemical rather than an parasite, virus, or some other type of pest killer.

Petroleum products — Anything which contains gasoline, kerosene, oil, or similar products.

Phytotoxicity — Injury to plant life caused by a chemical or other agent.

pH — The chemist's measure of acidity and alkalinity. It is a scale in which the figure 7 indicates neutral, figures below 7 indicate acidity, and figures above 7 indicate an alkaline reaction.

Photosynthesis — The process by which carbohydrates are manufactured by the chlorophyll-bearing cell granules (chloroplasts) from carbon dioxide and water by exposure to the energy of sunlight.

Point of drip or runoff — When a spray is applied until it starts to run or drip off the ends of the leaves and down the stems of plants or off the hair or feathers of animals.

Poison — Any chemical or agent that can cause illness or death when eaten, absorbed through the skin, inhaled, or otherwise absorbed by man, animals, or plants.

Poison control center — An agency (usually a hospital) in all the major cities which is informed of the proper first aid and antidotes for poisoning emergencies — including pesticide poisonings.

Pollinators — Bees, flies, and other insects which visit flowers and carry pollen from flower to flower for many plants to produce fruit, vegetables, nuts, and seeds.

Pollute — To make unclean or unsafe through carelessness or misuse.

Port of entry — Place where foreign goods (plants, animals, crops, etc.) enter the U.S.

Postemergence — After young plants push up through the soil.

Product — The herbicide as it is sold commercially. It contains not only the active ingredients, but also various solvents, cosolvents, surfactants, carriers, and other adjuvants that are designated as inert ingredients.

Propagule(s) — Viable seeds and/or living, vegetatively (nonsexual) formed plant parts which can originate new plants under the proper conditions and which are often disseminated by various environmental and/or biological buds, free of the source plant, which can originate one complete, new plant during the following growing season as in some Potamogeton species.

Preemergence — The time period between planting seeds and the seedlings pushing up through the soil.

Preharvest — The time period just before a crop is ready to be picked, cut or dug.

Pressure — The amount of force on a certain area. The pressure of a liquid pesticide forced out of a nozzle to form a spray is measured in pounds per square inch.

Product — A term used to describe the pesticide as it is sold — it usually contains the pesticide chemical plus a number of additives.

Proprietary mixture — One that is commercially available.

Protectant — A pesticide applied before pests are actually found, but where they are expected. The pests are destroyed before they cause any disease, damage, or harm.

Protective gear — Clothes and equipment that guard a person against injury or death when using poisonous pesticides. They would include gloves, apron, shoes, coveralls, hat, cartridge, respirator, and gas mask.

psi — Pounds per square inch.

Rate — The amount of a material which is being delivered to a plant, animal or surface. Usually measured a per acre, per 1000 ft^2 or per hour.

Recommended dosage — Advice from a County Agency, Extension Specialist, or other authority or written on the label on how much of a pesticide to use for preventing damage by or destroying a pest. This amount is not always the maximum allowed by law.

Registration — Approval by the Environmental Protection Agency of a pesticide for uses as stated on its label.

Regulatory officials — Those persons working for the federal or state government who enforce the rules and laws.

Repellent — A pesticide that keeps or drives insects or other pests away from the plant, animal or surface treated.

Residual pesticide — a pesticide that can destroy pests or keep them from causing damage for long periods of time after it is applied (days, weeks, months).

Residue — The amount of pesticide that remains on a crop, animal, or surface for a while after it has been treated. Not the same as deposit.

Resistant species — One that is difficult to kill; the use of the herbicide is not recommended.

Respiratory — A face mask which filters out poisonous gases and particles from the air so that a person can breathe and work safely.

Restrictions — Limitations.

rpm — Revolutions per minute.

Rodenticide — A pesticide used to control rats, mice, rabbits, and their relatives.

Runaway pests — Insects, diseases, weeds, or other pests which get into an area for the first time and therefore have no natural enemies — they often reproduce in large numbers and overrun an area.

Scientific name — The one name used throughout the world by scientists for each plant and animal. The names are based on the Latin or Greek languages.

Seizure — To take or impound a crop or animal if it contains more than the allowable pesticide residue.

Selective pesticide, specific pesticide — A pesticide which will control only a few pest species and is not as poisonous to other plants and animals.

Sensitive areas — Places where pesticides could cause great harm if not used with special care and caution. Examples: houses, barns, parks, ponds, streams, etc.

Sensitive crops — Crops which are easily injured by pesticide chemicals — even slight drift could cause a lot of damage.

Shock — The severe reaction of the human body to a serious injury which can result in death if not treated, even if the injury itself would not.

Short-term pesticide — A pesticide which breaks down almost immediately after application into nontoxic by-products.

Signal word — Word which must appear on pesticide labels to show how toxic the pesticide is. The signal words used are "Danger-Poison", "Warning", or "Caution".

Slurry — A water mixture or suspension of an insoluble herbicide.

Species (plural: species) — A morphologically, genetically, and ecologically defined biological entity to which a binomial and authority is given; e.g., *Potamogeton filiformis* Pers., the slender-leaf Potamogeton.

Solvent — A liquid, such as water, oil, or kerosene, used to dissolve other materials such as herbicides.

Spot treatment — Application of an herbicide to individual plants or small clumps of plants.

Spreader-sticker — A surfactant closely related to wetting agents that facilitates spreading and increases sticking of an herbicide on vegetation.

Standing crop — The quantity of living organisms present in an environment at a selected point in time.

Stem-foliage application — An application of an herbicide to both stems and leaves of a plant.

Sterilize — Treat with a chemical or other agent to kill every living thing in a certain area.

Suction hoses — The hose through which water is pulled from a pond or stream, or spray from the spray tank to the pump.

Surface spray — A pesticide spray which is applied in order to completely cover the entire outside of the object to be protected.

Surface water — Rivers, lakes, ponds, streams, etc. which are located above ground.

Surfactant — A chemical or agent used in a pesticide formulation to make mixing easier and help the material to spread over and completely wet the surface to be sprayed. Examples: detergent, emulsifier, and wetting agent.

Susceptible — Can be killed or injured by the pesticide at the rates used.

Suspended — A pesticide use that is no longer legal and remaining stocks cannot be used. More severe than canceled.

Swath — The width of the area covered by a sprayer making one sweep or one trip across the field or other treated area.

Sympton — A warning that something is wrong. An outward signal of a disease or poisoning in a plant or animal or man.

Systemic — A pesticide that is taken up by one part of a plant or animal and moved to another section where it acts against a pest.

Target — The area, buildings, plants, animals, or pests intended to be treated with a pesticide application.

Taxon (plural: taxa) — A unit of plant classification of any chosen rank; e.g., Potamogetonaceae (the Potamogeton family). *Potamogeton* (everything included in this genus), *Potamogeton foliosus* Raf. (everything included in this species), or *Potamogeton foliosus* Raf. variety *macellus* Fern. (everything included in this variety of *Potamogeton foliosus* Raf.).

Taxonomic authority — Biologist who described the morphological, ecological, and genetic limits of any taxon or who more recently revised the taxonomic position of a taxon.

Technical material or pesticide — The pesticide as it is first manufactured by the company before formulation. It is usually almost pure.

Test animals — Laboratory animals, usually rats, fish, birds, mice, or rabbits used to determine the toxicity and hazards of different pesticides.

Thermal — Related to heat.

Tolerance — The amount of a pesticide that can remain on any food (plant or animal) that is to be eaten by livestock or humans. The tolerance is set by the EPA.

Tolerant — Not susceptible to (injured by) a pesticide application.

Total alkalinity — Total of hydroxice (OH^-), carbonate ($CO_3^{=}$), and biocarbonate (HCO_3^-) ions in water; the capability of water to accept protons.

Toxic — Poisonous, deadly, injurious to plants, animals, or humans.

Toxicant — A poison. The chemical in a pesticide formulation that can injure or kill the pest as well as humans, animals, or plants.

Toxicity — "How poisonous" a pesticide is to a living organism.

Trade name — A brand name. The name given to a pesticide by a manufacturing company to identify it as their product.

Translocated herbicide — One that is moved within the plant from the point of entry.

Transport — Carry from one place to another — usually in a car or truck.

Treated area — A building, field, forest, garden, or other place where a pesticide is applied.

Ultra-Low Volume (ULV) — The application of a pesticide that is almost pure toxicant or technical material by spraying it in extremely small amounts over a large area — usually only a few ounces per acre.

Unauthorized persons — People who have no right doing something because they have not been told or trained to do it.

Uncontaminated — Does not contain hazardous pesticide residues.

Underground water — Waterways which are located beneath the soil surface which wells get their water from.

Uninformed persons — People who are not trained to use and handle pesticide safely.

Uniformly — Done exactly the same way each time or over each area. Done evenly.

Vapor drift — The movement of herbicidal vapors from the area of application.

Viscosity of oil — Expressed in time (seconds) required for 60 cc of heated oil to flow through a Saybolt Universal Viscosimeter.

Volatile — A compound is volatile when it evaporates or vaporizes (changes from a liquid to a gas) at ordinary temperatures on exposure to the air.

Water hardness — The total concentration of calcium, magnesium, and the metallic ions that have the capacity to precipitate soap (soft water = 50 mg/ℓ, hard water = 150 mg/ℓ).

Water-soluble power — A finely ground powder which will dissolve in water.

Weed — A plant growing where it is not desired.

Weed eradication — The complete elimination of all live plants, plant parts, and seeds of a weed infestation from an area.

Wetland plants — Terrestrial plants which preferentially grow in 100% water-saturated soil.

Wettable powder — A finely ground powder plus a wetting agent to keep the particles suspended in but not floating on the water to which it is added.

Wetting agent — A compound that when added to a spray causes the spray to contact plant surfaces more thoroughly.

WSSA — Weed Science Society of America.

Index

INDEX

A

B

C

D

E

L

M

N

O

P

Q

R

S

T